Books By Wilford H. Welch

The Tactics of Hope
How Social Entrepreneurs are Changing our World

In Our Hands
A Handbook for Intergenerational Actions to Solve the Climate Crisis

Values and Circumstances that Shaped a Life

Forthcoming Book

Your Choice
We have the technologies to Solve the Climate Crisis
We Now Need All of Us to Do Our Part

Values
&
Circumstances
That Shaped A
Life

A Wild Journey

By Wilford H. Welch

ISBN - Paperback: 979-8-9991428-1-8

First Edition: July 2025

I dedicate this book to all my family members

The Welches: *Father, Joy, Harry, Betsy, Kim, David, Toby, Peg, Barclay, Kristen, Bonnie, Gary, Christian, Noble, Suki, Alison, Will, Jake, Whitney, Hilary, Dave, Taylor, Morgan, Phoebe, Dave, Lindsey, Perry, Cathie, Perry Jr., Karen, Reed, Catie, Kelsey, Andrew, Mariann, Zander, Liam, Carolyn, Topher, Natalie, John, Lili, Wendy, Eric, Kristen, Nick, Peter; Del, Carole, Ashley, John, Noelle, Aydan, Shandy, Hans, Finley, Ayla, Carolyn, Alex, Paul, Hadley, Luke, Sam, Nell*

The Angermeirs: *Carole with her daughter Teresa, (left), her son Charlie (right) and her three grandchildren Ashley, Sophia and Kennedy.* **Not shown;** *Carole's father Paul, her mother Margaret, her brother Stan, his wife Anne and great grandchildren Coralie and Jackson, and their father Joe. Also, Charlie's wife 'Stanton' and her daughter Astrid*

Table of Contents

Preface

The circumstances we are born into and the values each of us has embraced over our lifetimes have long been of interest to me. This book is my opportunity to explore these questions in some depth and to share with others what I have learned.

I have written this book with several objectives in mind:

- I want to explore how the circumstances in which I was born and the values I have embraced over the past 86 years have shaped the choices I have made along the way.

- To explore the many phases of my life.

- To share the analytical methods I have been using every day for over fifty years to make sense of what is happening around the world, including Driving Forces Analysis, the development of Alternative Futures, and Scenario Planning.

- To share my observations about where I believe humanity is heading and the challenges we are facing.

I am not suggesting that the values I express here are for you to embrace but may cause you to reflect on your own.

The Values That Guide Us

The values we hold seem like invisible threads that are woven into the tapestry that is one's life story. And, much like invisible ink can uncover messages buried below the surface, this book seeks to uncover many of the values, as well as the circumstances, that have driven the choices I have made.

Values:

A person's principles or standards of behavior.

One's judgment of what is important in life.

The Circumstances That Shape Us

We have a thirty-three-year-old man we consider a son named Kelly Yadessa. His mother was a relatively poor Ethiopian woman. His father was an Italian construction worker. For years, Kelly was one of those children you might

encounter in developing countries who, as he told me, might pull on your shirt, asking if he might practice his English.

One day in Ethiopia seven years ago, Kelly was called by a friend in the Ethiopian security service who told him that he and thirty-two others in his protest movement seeking to protect the tribes in the South from draconian Ethiopian government policies were about to be arrested. Kelly's friend told him that he must flee the country that night. I had already helped Kelly get a visa to come to the U.S. at some time to promote his new travel company, so he had the means to flee the country on short notice.

The next day, Carole and I got a call from Kelly from the San Francisco airport saying, "Hi, Mama and Papa, I am here! Can I come and stay with you?" – which he did for the next six months.

After five years in the San Francisco Bay area, Kelly was overseeing 250 security workers in buildings throughout the Bay area. The other 32 members of his protest movement have not been heard from again.

Kelly, Carole, and me, greeting Kelly's wife Tsega and their two boys. Michael and young Wilford upon their arrival at San Francisco airport after six years apart

What were the motivations that caused Kelly to make the very significant choices in his life, and where did they come from? If you had an opportunity to ask him, you might be surprised to learn how his values, largely influenced by his belief that "God has a plan for me," appear to guide him in all his decisions.

Carole and I recently went to a lecture to hear the story of Myron Spaulding's life. Myron was born in 1905 to an impoverished family in Northern California. Soon after his birth, his mother abandoned him, and soon thereafter, his father also abandoned him but gave him a violin. After many

years in orphanages and foster homes, Myron became very passionate about the violin and ultimately became the first violinist in a number of orchestras, including the orchestra in San Francisco.

Myron Spaulding playing first Violin in the San Francisco Symphony

After twenty-two years as a first violinist, Myron followed his other passion — boat design and boat building — to become one of the great sailboat designers and builders on the West Coast.

Myron Spaulding on the San Francisco Bay

The circumstances into which Kelly and Myron were born were far different from mine. And the values they acquired along the way were also probably very different. But their values must have played major roles in shaping the choices they made and the extraordinary lives they had. We can only speculate.

Chapter One

MY EARLY LIFE IN THE COUNTRY

I was not brought up in the challenging circumstances that Kelly and Myron had to navigate. Rather, I was born in the country outside New Haven, Connecticut, in 1938, in an upper-middle-class family two months before WWII started with Hitler's invasion of Poland.

Joy at a Yale football game

My family was a large one, made up of my father, G. Harold Welch, and my mother, Harriet Wilford Hitchcock Welch, who I always called "Joy," and for good reason. I had three older brothers, Harry, Noble, and Perry, followed by my sister Carolyn, who was born in January of 1945, shortly before the war ended. I vividly remember hearing on the radio in April of 1945 that the U.S. had just dropped an atomic bomb on Hiroshima. That was followed on September 2nd by Japan's surrender, ending a dark period in humanity's journey during which 70-85 million soldiers and civilians died on the battlefield or of war-related diseases or famines.

As I look back, I also realize that I have lived during years of relative peace and prosperity. When the war ended when I was seven, the United States was left intact while

1

much of the rest of the world was in ruins. Our industries, which had focused on war production, quickly shifted to providing the American people with what they were yearning for: new cars to drive, new houses to live in, and all the consumer goods they had been deprived of during the war. As a result, my life has been one of relative safety and comfort and filled with opportunities. I had yet to learn how fortunate I was in comparison to most people in the United States and around the world.

Top: Joy, Carolyn, Harry, Perry, father Bottom: Noble, Wilford and our golden retriever

The extraordinary trips we took as a family to faraway destinations and the opportunities we had to get excellent educations were but a few of the things we were fortunate to experience.

Clearly, I was also born with a "positive gene" in terms of my attitude toward life, and that attitude has influenced how I perceive the world and the decisions I have made. My wife Carole has gone so far as to suggest that she plans to have inscribed on my grave marker: *"Wilford H. Welch, Yes to Life."* Clearly, I was also born into "privilege," by which I mean a stable, loving family, a life in the country filled with nature and animals, comfortable financial means, and opportunities to have an excellent education.

"Christmas Tree Hill" where we were brought up, with "Sleeping Giant Mountain" behind it

I clearly remember one day when I was six or so when we all went down to the pond with rakes and waded in where the rowboat can be seen in the photo above. We wanted to drag the vegetation in the pond to the other end of the shore and burn it. During skating season, we wanted to be able to skate and play hockey on "black ice" without vegetation coming up from the bottom and mixing with the ice.

I raked my load to the other end of the pond (left of the tallest tree in the foreground) and stood on what I thought was a rock. I then got under the "rock" with my rake and heaved it onto land, only to find that it was a snapping turtle about eighteen inches in diameter! His powerful mouth could have caused serious damage.

My father, who always wanted to have us call him "Father dear," was an up-and-coming banker in New Haven, Connecticut, who strove to give his children the very best in terms of values, education, and opportunities. When I was seven to thirteen years old, I went to Foote School in New Haven, and when he would drive me to school, he often had me recite this list of thirteen things he expected of his children. (See on next page.)

He asked me to call him "Father dear" and recite the thirteen things he expected of each of his children, which seems quaint and overly patriarchal today. But my father, who was born in 1897, was brought up at a time when fathers were expected to be the disciplinarians and the ones whose primary job was to earn the income needed, while their wives were expected to stay at home and take care of the children.

Outline of what father expects the boys to do:

Always tell the truth--which means to tell exactly what happened, under all circumstances.

Be a good sport--which means to play fairly, to always try to win, but in the event that you lose, congratulate the winner and strive to win the next time.

Be a perfect gentleman--which means to be polite and kind to everyone.

Do first things first.

Things worth doing are worth doing well.

Do one thing at a time and do it well.

Keep hands off other people.

Do unto others as you would be done by.

Father is not interested in excuses, he is interested in results.

Always be a happy boy.

IMPORTANT--on your honor--which means to do the things that you are expected to do on all occasions, regardless of whether or not father or mother are with you.

Do things for other people, and do a little more than is expected of you.

Mind quickly, and then you will be happy boys with many friends and you will have many surprises.

The only jobs most women could get outside the home at that time were to become teachers, nurses, or secretaries. Women could not even apply for a bank loan without their husband's signature until 1974 when the law was changed. "Joy" devoted herself to her children and her husband's well-being while also being very active in community organizations and events. For example, Father and Joy were very involved in the first Hospice to be established in the United States, in Branford, Connecticut, the town where my mother was born.

When not at school during those years, I was mostly working on our property. That included mowing the fields with our old Ford tractor and tending to our large garden, which we turned into a "Victory Garden" during the war. It also included my cutting down trees, sawing them into logs for the fireplace, and taking care of our animals. I had very few friends, as we were quite isolated, and I could not drive.

Every morning and evening, I had to "do the animals," which meant watering and feeding our two horses and "Peggy," our Shetland pony, letting them out into the pasture and mucking out their stalls. Next, I would carry two large pails of water the one hundred yards around the garden to our two chicken coops, put grain in their feeding troughs, collect the eggs, and then let the chickens out into their fenced yard. One snowy morning, I opened the chicken coop door, and a red fox ran out past me. I chased him, did a flying tackle, and grabbed his tail. Fortunately, his tail slipped through my hands, and I did not end up with a badly scarred face for the rest of my life.

My brothers Perry and Noble are seen here riding on "Peggy," our wonderful companion while we were growing up.

Peggy's gravestone which I carried back from Massachusetts in 2010 and put on our roof deck

Killing the old roosters and hens was also one of our tasks. I had not recalled that my sister Carolyn did the killing as well as her brothers until very recently when I found this account that she had written:

It always seemed a bit daunting. First, I fell in love with each baby chick when they were brought home in their octagonal cardboard box each spring and kept under a warm light in our basement. Once they were sturdy enough and had turned from fluffy yellow to white or brown, they were taken to the chicken coop for quite a different existence. Many years later, I can still smell their chicken poop!

It was my daily chore to carry two large metal buckets of water from the barn to the coops in the morning and evening, just as my brothers had done before me. I remember wondering more than once where the heck those big strapping brothers of mine were when I needed them.

Once a year, the time would come that I dreaded most: Killing some chickens for us to eat. The scene: Hot steam rising from a big metal bucket over a small wood fire behind the barn, the chopping block and axe not far away. I used a long, bent wire with a hook on the end to snare each chicken as they ran around, trying to evade my hook. Next, I would tie their legs together and carry them upside down to the "guillotine." Laying their long necks perfectly on the chopping block was an art. I had to stretch their necks on the block and then act quickly so the kill would be swift and clean. The sound will forever stay in my ears. A loud "thwack," and the deed was done.

I had never seen so much blood in my life. Their lifeless

bodies would gush blood from what remained of the neck and body. The oddest sight of all was to see the dead chickens still jumping even though their heads had been cut off.

My next task was to dunk each chicken's body three or four times into the bucket of boiling water so that I could easily pluck off their feathers. Each stage seemed gruesome, but the final stage was the most sobering: cutting a slit between the chicken's legs, reaching inside, and pulling out the liver, heart, and kidney. All the chickens were then transported to our basement, where my dear and wonderful mother and I would spend a full day packaging each chicken in plastic wrap ready to be frozen and later cooked for our many family dinners at "Christmas Tree Hill."

During the summer of my thirteenth year, Carson Nile was my best friend. He lived in a house nearby and took care of our property. He was in his forties and very experienced in the backwoods. The two of us spent the entire summer turning a large, forested area into a pasture for the horses. We first cut down the trees, most of them a good 50 feet tall. We then trimmed off all the branches and dragged them to "burn piles."

Then came the best part: digging around each of their large roots and putting sticks of dynamite under each to blow the roots up. We cut each stick of dynamite in half and then cut each "wick" (the chord which burns quickly when lit to detonate each stick of dynamite) into different lengths, so if we ignited them in the correct order, they would all go off at approximately the same time. We then attached a detonating cap to each wick.

Next, we tucked each stick of dynamite under a root. Carson would shout to me to light the longest "wick" and then the other four in order of length before we ran like hell and jumped into our "fox holes," just like the soldiers had done during the war. It was the most exciting experience of my life up to that time.

Our next task was to cut the downed trees into manageable lengths for our fireplaces and use the tractor and chain to haul the downed tree trunks, roots, and branches to the burn piles. It was undoubtedly during that summer that I developed my love of outdoor fires and tractors, which continues to this day.

In addition to wanting their children to grow up to be good citizens, Father and Joy were committed to making sure we got a great education. They also wanted us to travel the world, where, for the first time, I came to realize how fortunate I was in comparison to most people. My first trip abroad was a family trip to Cuba in 1954, shortly after Fidel Castro had taken over the country. Other trips with Father, and some with Joy, were to Sweden, Ghana, Nigeria, South Africa, China, Mongolia, and the U.S.S.R. After each trip, Father would write up where we had gone, what we had done, and what we had learned. He later put them all into a book entitled "Tripping with Father."

As I look back on those years, the values instilled in me by my parents and the circumstances I was privileged to have as a young boy are clear: the importance of a loving, stable family, the value of hard work, an appreciation of the natural world; the value of being exposed to other cultures, their histories, and aspirations.

Chapter Two

GROTON SCHOOL

WHERE VALUES WERE FIRST & FOREMOST

Each of my three older brothers had gone to Groton School, a private school in Massachusetts founded in 1884 by an Episcopalian minister, Endicott Peabody. Every student was required to attend chapel each morning and twice on Sunday. It was assumed that I would go to Groton, too. I took the entrance exam and did ok but was turned down. That delivered a serious blow to my young sense of self, and I started to question my competence.

My father then drove me up to visit St. Mark's school, and on the way north from New Haven on Highway 95, I looked to the right and saw a sign reading, "Fine for throwing trash." Confused, I asked my father, "Why would they put up a sign saying it was "fine to throw trash?" I will never forget the look on my father's face or my next thought, which essentially was that I really was not very smart and would not have the successful life I presumed each of my brothers would have.

I ultimately did get into Groton, and for the next six years, with great determination, I took on every challenge I could to prove to myself and others that I was smart and capable. I was my class president all six years and head of school my sixth form year; I was captain of our football team and also played on our varsity hockey and baseball teams.

I was so determined that I talked our football coach, Jack Davison, into letting me kick the extra point after the very tough "townies" at English High had tied the score at 13-13 late in the game. He reluctantly agreed, and so, despite my having seriously injured my left ankle in the third quarter, I kicked the football through the goalposts, and we won the game 14-13.

Groton School Campus

The next day, I was driven to a hospital in Boston and came back to school with a cast from my toes to my hip that I had to wear for the next eight weeks. Our football team was undefeated in our junior year and only lost one game in our senior year. The hockey team I played in my senior year went undefeated, with the captain of our hockey team, Ken Maclean, later becoming the Yale hockey captain and our center, Stewart Forbes, becoming captain of the Harvard hockey team. Not bad for a class of only forty-three boys.

I loved Groton, and Groton seemed to love me back. Groton was an all-boys school at the time, and in retrospect, I feel that not having the distraction of girls on campus was good for me. Yale was also all-male during my years there, but I do not feel that was wise as I feel it was important during those years to get to know the opposite sex.

Groton's 1956 football team. I am number 43

We had a great class, and we have stayed in touch over the years. In addition to my classmates mentioned towards the end of this book, I'd like to mention John Bingham and John Carmody, who were with me all six years at Groton and the next four at Yale. Nat Coolidge went on to Harvard.

FALL SPORTS

On crutches from the leg injury against English High

In the summer of 1957, Gordon Gund, Sam Webb, and I drove a Pontiac station wagon across the country and back. Here is a photo of the three of us thinking (incorrectly) that we were very cool.

Groton instilled in me such values as the importance of teamwork and leadership, the power of fierce determination, the value of deep friendships, and the power of a positive attitude. And finally, the importance of being of service to others. Groton's motto in Latin is "Que Sevire Est Regnari," which, loosely translated, means "To serve is to be free." Being of service was regularly impressed upon us.

Gordon Gund, Sam Webb and me on our 1957 round trip coast to coast trip

In 1954, while at Groton, my father took my brother Perry and me to Cabo Blanco along the coast of Peru to fish for marlin. I was in the chair when a black marlin was spotted off to our starboard side a mile away, its prominent fin cutting through the water. The captain had the bait pass a quarter of a mile in front of the marlin. He went for it, and the fight was on. Once brought onboard and to the dock to be weighed, we found that I had caught a 770-pound, 12-foot-8-inch-long black marlin, which landed me in the Guinness Book of Records for a while. And, while that was a great thrill, I now realize, after a lot of trout and salmon fishing, that I enjoy even more than fishing in the ocean, fishing in the rivers of Maine, Wyoming, and elsewhere in the natural world. Here I am with a fish I caught in Wyoming.

Father and me with the back marlin

Chapter Three

THE UNDEFEATED, UNTIED, YALE FOOTBALL TEAM OF 1960

I went on to Yale University in the fall of 1957, where I stayed with Sam Web and Nick Gardiner and majored in economics and history. I was an adequate student but did not have a passion for academic learning. And, while those who knew me seemed to think of me as quite successful, I had an uncomfortable feeling that I was a bit of an imposter. I do not know where that uncomfortable feeling came from, and I have never commented on this before writing this book. It was much the same feeling I had as a result of the "fine for throwing trash" incident six years earlier.

In what I now see as a pattern, I sought to overcome this uncomfortable feeling by playing football all four years. I was the quarterback of our junior varsity team and was one of the backup quarterbacks on our undefeated, untied team of 1960, and I got my letter. I was also the quarterback who played against our first-string defense during practices each week. I weighed only 172 pounds and was not very fast, but I seem to have made up for this by again showing a lot of grit. Our team remains the only undefeated, untied team in Yale's history since 1923, more than 100 years.

The Yale undefeated, untied Football Team of 1960 (I am number 14)

Six of our players were named to the first, second, or third-string All-American football teams in the United States. Our captain, Mike Pyle, was drafted by the Chicago Bears, and he became captain and center for the next ten years. Our first-string quarterback, Tom Singleton, went on to become the first-string quarterback for the U.S. Marine Corps.

1960 Yale Bulldogs football

Ivy League champion
co-Eastern champion

Conference	Ivy League
Ranking	
Coaches	No. 18
AP	No. 14
Record	9–0 (7–0 Ivy)
Head coach	Jordan Olivar (9th season)
Captain	Mike Pyle[1]
Home stadium	Yale Bowl

During regular games, I seem to have developed a modest fan base that could occasionally be heard when we were far enough ahead in any given game: "We want Welch, We want Welch." I also played indoor polo all four years, during which our team won the U.S. intercollegiate championship, a satisfying but rather modest accomplishment. I accepted an invitation to join Berzelius, one of Yale's secret societies, in my final year, where I was elected as president. To my delight, fifty years after we graduated, my classmate and Berzelius delegation member, Joe Novitski, happened to buy a houseboat on our pier about three hundred feet down our dock. We see Joe and his partner, Susan, almost daily, and enjoyably so.

Upon reflection, I realize that I was so involved in life at the university that I was not really aware of the social issues that were building up in my country, including the vast inequities among our citizens, the lack of many African Americans and Jews in my class, the growing Civil Rights Movement and the teachings of Martin Luther King. All that came into focus when I was studying law at Berkeley three years later.

And, yet again, Yale reinforced one of the values instilled in me at Groton, that working together as a team, on or off the field, had become central to my makeup. Curiosity would soon follow.

Chapter Four

BEGINNING TO EXPLORE THE WORLD, PARTICULARLY CHINA

During spring break of my senior year at Yale, my father and I went to Sweden, where I was smitten by a beautiful eighteen-year-old girl. (Keep in mind that after six years at an all-boys prep school and four years at an all-boys university, I was likely to find almost any member of the opposite sex attractive).

At the end of the trip, I decided to go back to Sweden that summer and pursue her. I got a summer job at Volvo to help finance my chase. And while I did not get the girl, I got something far more important, which would set the course for the rest of my life:

One Sunday morning in my apartment in Sweden, I panicked over what I might do as a professional after graduation. All I was sure of was that I wanted to work internationally and did not want to go into banking as my father and my three older brothers had chosen to do.

In the midst of my panic, I took out a piece of paper and wrote down this question: "What are some major events that are likely to occur internationally during my most productive years that I might hook my star to?"

This response came out of my unconscious within a minute or two, and I wrote it down: "The world's most powerful country, the United States, and the world's most populous country, the People's Republic of China, have to reestablish diplomatic relations." I reasoned that it would be critically important that the U.S. and China have channels established to be able to talk with each other about all sorts of things, from trade opportunities and barriers to the threat of nuclear war. At that point in my life I had not been to Asia, had taken no courses on China, or studied the language.

A couple of minutes later, a second insight emerged: I recalled that at the age of fourteen, while at Groton, I had been drawn to watch the "McCarthy Hearings" on television, during which Senator Joseph McCarthy, of Wisconsin, accused the "Old China Hands" in the U.S. diplomatic service of "Giving China away." McCarthy was all about power and seemed to have no concern for the lives he destroyed along the way. (Sound familiar?) Specifically, he destroyed the careers of many "Old China Hands," along with others in our society, particularly many in Hollywood, who he accused of being "card-carrying communists." I saw an opportunity for a new group of Chinese diplomats to emerge and decided then and there in Sweden that Sunday morning that I wanted to play a role in the re-establishment of diplomatic relations between the United States and the People's Republic of China.

As soon as I returned to Yale for my senior year, I started to put my unexpected goal into reality. I got a fellowship from the Yale-China Association to go to Hong Kong for two years to teach English as a second language (ESL). I would be teaching Chinese refugees from mainland China at the New Asia College in Hong Kong.

Shortly after graduation from Yale, I flew off to Asia to start a totally new phase of my life. I was driven by a great curiosity about Asia and its many cultures, particularly China. I also learned, for the first time, that I loved to teach.

Yale classmates Tom Davenport, Greg Prince and me in a Chinese language newspaper

The Chinese that is spoken in Hong Kong is predominantly Cantonese, which was a problem if I wanted to speak good Mandarin without a Cantonese accent. I therefore sought out Margaret Chai, the leading Mandarin teacher at Hong Kong University, and moved into the rental apartment she had in Kowloon, a part of Hong Kong. We agreed to speak only Mandarin to each other to the extent I could keep up.

One Sunday, Margaret showed me this Japanese cavalry banner, which she said was taken as a war trophy by the leader of a Chinese cavalry that captured a Japanese cavalry troop late in the Second World War. I asked her many questions about it but never said how excited I was to see it. Shortly after that, I left Hong Kong and thought nothing more about it.

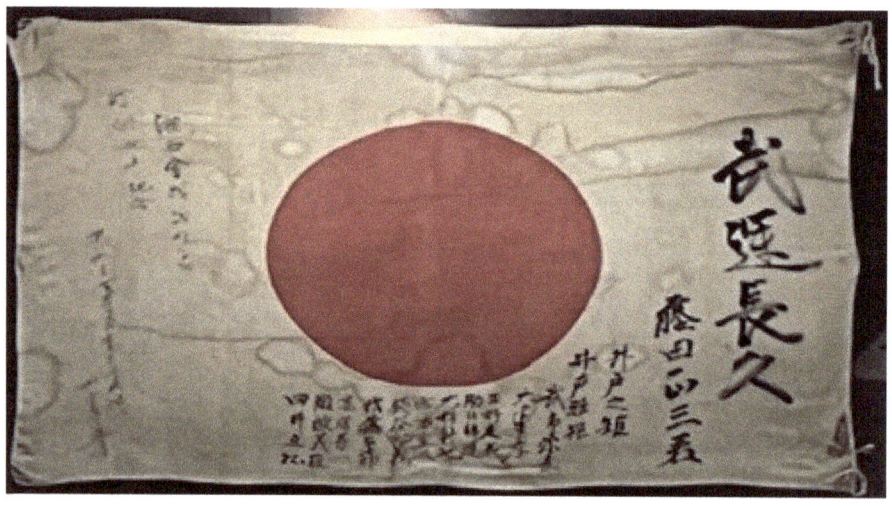

Japanese cavalry banner captured by a Chinese cavalry

Three years later, Margaret called me while I was at Berkeley studying the Chinese legal system to tell me that she had moved to California to teach Mandarin at the Monterey Language Institute. She suggested that we have dinner. At the end of dinner, she reached into a paper bag, pulled out the banner, and said: "You told me in Hong Kong that you were committed to bringing about harmony between the peoples of the United States and the peoples of China. I see that you have kept your word. I would like to give this banner to you as a token of gratitude from the Chinese people." Her comment and her gift mean a great deal to me.

During my two summers in Hong Kong, I went to Taiwan to study Mandarin, and when back in the U.S., I studied advanced Chinese reading and developed a reading vocabulary of about 3,000 characters, the rough number of characters needed to read a newspaper.

Every other time there was a break from teaching in

With members of the Meo tribe in northern Thailand

Hong Kong, I headed off to explore the countries of Southeast Asia, including South Vietnam, Cambodia, Laos, Thailand, Malaysia, Singapore, the Philippines, and Indonesia. These countries had not yet embraced Western ways, such as our tall skyscrapers and shopping malls, and I was exposed to the richness of their different histories and cultures. I took many trips to Vietnam in particular because I had good friends there, namely Jeff Farrell and Peter Glick, who were working for non-profits. One of them was also working for the CIA.

I had four challenging encounters in Asia and elsewhere around the world during my Yale-China days and in later years:

The first incident occurred one night in Singapore in 1962. At that time, Singapore was quite a lawless place, with Chinese Triad Societies and communist-controlled labor unions creating chaos. An American friend and I went to a very local Chinese restaurant, where we ended up at a small, square table with two British sailors. My friend David sat to my left, a nice young sailor sat across from me, and his surly colleague sat to my right. My friend proceeded to get very drunk and began accusing the British government of not having the guts to remain in Singapore in Britain's post-colonial retreat "East of Suez."

After taking this for a few minutes, the surly sailor pulled out a stiletto knife and was ready to stab my very drunk and foolish friend. Having seen too many John Wayne movies, I instinctively lifted "our" end of the table over on the two Brits, thrust my right arm between my friend's legs, threw him on my back in a "fireman's carry," and raced out of the restaurant with the Brits presumably in hot pursuit. (I had no time to look back). I threw my friend into a taxi on top of a Chinese lady who was already in the back seat, and we sped off and out of harm's way.

The second incident occurred in Bolivia in 1973 when my wife, Del, and daughters, Ashley and Shandy, joined me when I was working in Bolivia for Arthur D. Little, Inc. We rented a small home twenty minutes from La Paz that backed right up against an army base.

One memory that has stuck with me was of the gardener who cut the grass in our small yard every week using the sharp edge of a soup can top.

I had two weeks off, and we decided to stay in Bolivia. Our initial plans were to hire Jorge Zallas, a local guide, to take Del and me down the Rio Beni River where Che Guevara, the Cuban revolutionary, had been assassinated in 1967, six years before. Jorge's plan was to drive 50 miles to the Rio Beni River, buy six 20' logs, tie them together, and raft thirty miles down the river.

Jorge assured me he knew what he was doing and that we would be safe. When I asked if anyone had been injured during his previous trips down the river, he admitted that two of his crew had died but added with a smile, "No one who had ever paid has died." With that comforting assurance, we decided to leave the next Wednesday for our Rio Beni adventure, only to be informed by Jorge that he had to cancel the trip. He was not keen to tell me why until finally he blurted out: "Because some of us are planning a coup!."

The next Wednesday, I was driven to our office only to receive a frantic call from Del, who shouted: "Planes are strafing the house! Come home as fast as possible." When I got home, Del, Ashley, and Shandy were safe but very shaken. Since Jorge had told me of the planned coup, it was not a stretch to assume that word had also gotten to the Bolivian President, Juan Jose Torres. President Torres was scheduled to preside over the graduation of Army cadets that day in the army camp adjacent to our house and would be captured by the army in a coup. Torres still had control of the air force and had sent the planes to thwart the army's plans. No shots were fired, but Del did see the eyes of the pilots as they swooped within a hundred feet over the army barracks and our home. That was the last I heard of Jorge Zallas.

My third challenging encounter was in Mongolia in 1982 when my father, my sister Carolyn, and I were there. After several days, we were scheduled to take a train to the Siberian border town of Erlyan and on to Irkutsk, Siberia, followed by a three-day ride across ten time zones via the trans-Siberian railway to Moscow.

As was customary, the Mongolian government had taken our passports upon arrival into the country and would give them back to us upon our departure. When the time came to get our passports back, they said they would only give them back when we gave the Vice Minister of Tourism $700. I refused, and in a subsequent meeting with the Vice Minister, I told him that I planned to publish an article in the *WorldPaper* where I was the publisher, noting the incident and featuring photos I had taken that would embarrass his government. It was, in fact, an empty threat as I had taken no such photos.

We ultimately did give him $700. When our train got to the Soviet border, a young Mongolian soldier entered our cabin and demanded the film in my camera. I refused. Not knowing what to do, the soldier left and turned the problem over to his commanding officer, who came into our cabin with his gun drawn, looked at me, held out his hand, and said, "Film!" My bluff called. I took a film canister out of my camera and gave it to him, and we continued on to Siberia. I clearly remember having a very strong drink immediately after he stomped out, undoubtedly feeling he had outsmarted the American.

My final challenging encounter also occurred in Mongolia in 2003. My wife, Carole, and I were leading a ten-day trip on horseback in a very rural area of Mongolia for her adventure travel company, Cross Cultural Journeys. Learning that our guide was training the herdsmen in one valley to play polo, I volunteered to help.

Nomads playing polo in rural Mongolia. *Source:* Yak Polo, Wikipedia

One afternoon, after riding for some time and hitting a polo ball, the horse decided he had had enough of these strange sticks and balls whipping past his head. He lowered his head, raised his back legs up very high, and threw me right between his ears. I landed very hard on my head, cracking my helmet and doing serious, painful damage to my neck.

Two agonizing days later, I was strapped into the front

Stabilizing my neck with a towel, cardboard, and packing tape

seat of a Land Rover in which we forded four streams before arriving at a very rural hospital. There, I was X-rayed, standing up between two nurses, one holding an X-ray camera a few feet in front of me, the other holding the X-ray film behind me. Minutes later, I was told that I "had broken my neck in two places."

It took three days for Carole to speak with the Travelex Insurance Company in London via the Iridium Satellite phone we brought with us and the GPS we had brought as well to initiate an evacuation. We were able to give the military helicopter dispatched to pick me up our exact coordinates. The next day, a very large red Russian transport helicopter arrived and took Carole and me the 2 ½ hours to Ulan Bator, the Mongolian capital. There, an ambulance met us and took us to the hospital, where a CT scan confirmed that I had two breaks in my neck.

Christopher Reeves, the American actor who played Superman in the movies, had recently died in a similar accident. He was thrown off a horse and broke his neck. As a result, Carole and I were focused like lasers on trying to save my life. Over the next few days in an old Russian-style hotel in Ulan Bator, I sat in a wingback chair and stared at a spot on the wall, trying not to move my neck for fear of cutting a nerve in my vertebrae and becoming paralyzed like Christopher Reeves.

Never during my years playing sports have I felt more supported and part of a committed team, as during those thirteen days in Mongolia after the fall when Carole did everything possible to save my life. Fortunately, when we arrived home, Dr. Sigurd Berven, a spinal cord surgeon at the UCSF Medical Center in San Francisco, read the MRI and told us that I was a very lucky fellow. He said I indeed had chipped two vertebrae, but on the outside of my neck, and that I would be ok but that I would "have rotator cuff problems the rest of my life." In 2024, I had major surgeries on both my shoulders to address the problem caused by that annoyed and the most annoying Mongolian horse.

Near the end of my years in Asia during the early 1960s, I was asked by the Reverend Sidney Lovett to return to the Yale-China office in New Haven to travel around the U.S. to speak about the Yale-China program at Yale Alumni Association gatherings. As luck would have it, my host in Washington D.C. was William P. Bundy, the U.S. Assistant Secretary of State for East Asia and, therefore, America's top diplomat regarding U.S. policies in Asia. I stayed with him, and his wife, Mary, got to know him a bit, and he had a chance to take his measure of me. That proved a most consequential coincidence in my quest to play a role in the re-establishment of diplomatic relations between the U.S. and China.

Another consequential coincidence occurred when I was speaking about the Yale-China program in the San Francisco Bay area. I heard of a Professor at the University of California Law School named Jerome A. Cohen, who was the first professor of law in the U.S. studying what the legal system in Communist China looked like and how it was practiced. While my initial connection with Bill Bundy was great, I realized that two years teaching English and studying the Chinese language in Hong Kong at a refugee college was not nearly enough experience to give me much credibility as a "China Hand." And I reasoned that studying the Chinese legal system at a prestigious university just might give me enough credentials to be taken seriously. Jerry Cohen was delighted to have his first student, and I was accepted at the Berkeley Law School, starting in the fall.

My learnings from my summer in Sweden and my years in Asia were substantial. I found that following my intuition could have great value. I learned that I liked to learn more from experiencing the world than studying the world in school. I learned that I had made a good decision that Sunday morning in Sweden, even though I still do not know where that thought about my playing a role in U.S.-China relations had come from. And I learned that I loved to teach. Finally, my conviction about teamwork that I had experienced at Groton and Yale was amplified by the teamwork Carole and I shared in Mongolia to save my life. This was not the teamwork called for in a sport to win a game or the teamwork called for in a business to improve efficiency and profitability. This was the exhilarating feeling one has when another person is deeply committed to your welfare and you to theirs.

Chapter Five

LAW SCHOOL AT THE UNIVERSITY OF CALIFORNIA, BERKELEY, TO STUDY THE CHINESE LEGAL SYSTEM

In the fall of 1963, I entered law school at Berkeley. In many respects, I felt out of place. My fellow students were keen to practice law and make lots of money. I had no such interests. I had to study American law so I could study Chinese law, a subject in which my fellow students had no interest. I also felt that my becoming an attorney was a good backup profession if all else failed. While I passed the American law exam following my three years at Berkeley, I have never practiced law, but nonetheless, I gained a valuable understanding of how our laws and regulations impact our society. To be fully transparent, I went to law school to prove to myself and others that I was very capable and that my background in Chinese law would distinguish me in U.S. diplomatic circles.

And, just as my Yale-China years had opened my eyes up to Asia, so Berkeley opened my eyes to the civil rights and anti-Vietnam war movements going on in the U.S. – as well as the "free love" and "free speech" movements." Berkeley in 1964 was where these movements took off and spread across the United States. Michel Tigar, by far the smartest student in my law school class, was the founder of SLATE, a student movement that morphed into the Free Speech Movement.

Being exposed to that activism was an eye-opener to me. On one occasion, over the course of three days, about 2,000 students surrounded a police car that held a student they had arrested. The students held the police car and its occupants hostage for thirty-two hours across from the U.C. Berkeley president's office, feeding their police captives and not harming them but also sending an activist message. My professor of torts, Robert Cole, taught me a great deal about dealing with clashing groups when he skillfully negotiated a truce between the activists and the university.

A side note to my years at Berkeley was my decision in my second year that I did not want to marry my good friend in Hong Kong, Dzou Hwei-Ling. I did not feel it right to just send a letter, so I decided to fly to Hong Kong to tell her personally. My problem was that I did not have the $1,400 needed for the round-trip flights. I called my close friend Sam Webb, and he agreed to lend me the money, so I went. When I returned to Berkeley, I put an ad in the paper offering Mandarin language lessons and soon earned $1,400, which I sent back to Sam.

What seems important to me about my time at Berkeley was the power of activism, which may have been the genesis of my willingness to not accept the status quo but challenge it if I thought that was called for. And confirmation that I did not want to be either a banker or an attorney.

Chapter Six

THE U.S. DIPLOMATIC SERVICE – SERIOUS BUSINESS

After I made the connection with the State Department's Assistant Secretary of State William P. Bundy in Washington while speaking about the Yale-China program, Bundy would occasionally reach out to me about my plans after law school. The interchange evolved into an invitation to come to Washington and work directly for him. Needless to say, I accepted. In the fall of 1966, after marrying Adele Merrill in Cambridge, Massachusetts, she and I headed off to Washington.

Bill Bundy was very smart, demonstrated by the fact that he and his brother McGeorge Bundy, who had been President Kennedy's and President Johnson's National Security Advisor, both passed the SAT exam without making a single mistake. During the Second World War, Bill was in the OSS, the U.S. intelligence agency that was the precursor to the CIA, where he led a team of nine cryptanalysts working at Bletchley Park in England to break the German "Enigma code." He also had gone to both Groton and Yale, which clearly helped me. And I assume that he checked me out with a number of his friends in both institutions. Because I was in his office every day, I met on a regular basis with all the U.S. ambassadors to the Asian countries who reported to him, from Korea in the North to Australia in the South and all countries in between, including Vietnam and Taiwan. (The U.S. and China did not have diplomatic relations at that time, so no American could go to the People's Republic of China). I also interacted with many of the people in Washington and in Asia who were responsible for establishing U.S. economic and political policies in the region.

During my years with Bundy and my years with his successor, Ambassador Marshall Green, in the Nixon administration, I traveled to Vietnam on a number of occasions. The war was heating up, and the U.S. was getting mired deeper and deeper into a very unpopular war.

I felt I had a unique, three-sided perspective about Vietnam: 1) As a result of my trips to Vietnam while teaching in Hong Kong. 2) my exposure to the anti-Vietnam war protests at Berkeley, and 3) my now working within the U.S. government where much of U.S. Vietnam policy was being made, with Bundy being one of its architects. That said, President Johnson, a Texan and red-blooded American, had stated publicly that he would win the war, and others in our government engaged in Vietnam, such as Bundy, were expected to fall in line or get out of the way.

William P. Bundy, the U. S. top diplomat for East Asian Affairs

Bundy was often invited to speak on college campuses, where he knew he would be jeered and not likely to change any student minds, so he declined.

I volunteered on occasion to go for him, arguing that if we were in Vietnam and had a policy that so many Americans disagreed with, we had an obligation to listen to their views and share the official U.S. policy perspective. I had a few oranges thrown my way, but I very much appreciated the opportunity to do this. My best times were staying up until 1:00 AM or so, drinking beer with the student leaders who were against the war.

Late last night, while going through some of my fifty-year-old diplomatic files, I ran across a letter to Bundy from a professor at Kansas State College who had set up a "Teach-in" of students opposed to the war. The professor suggested to Bundy that I had accomplished what I was hoping. He wrote: "Mr. Welch was extremely well informed and effective in presenting his viewpoint and giving out relevant information. His ability to find broad areas of agreement between his position and those who opposed was remarkably constructive."

Of course, if I was to go out and speak officially about the war, I had to represent the official U.S. government position. That was tricky in that I was also questioning how the U.S. was prosecuting the war, not whether we should have gotten into Vietnam in the first place. I increasingly had come to the conclusion that our strategy of using our substantial military might was bound to fail against a guerrilla army fighting for their own country. This was much like the Algerians who had fought successfully against the French in the 1950s when the French tried to reassert control over the country after the Second World War in the face of rising anti-colonialism. This was much like the French after World War II, who tried to regain control of Vietnam but were defeated at the Battle of Dien Bien Phu in 1954.

I still felt the "West" had to 'hold the line" someplace in Asia in the midst of serious efforts by Soviet Russia and the People's Republic of China to take over countries that the colonial powers were retreating from all over the world. China had been taken over by the Communists in 1949/1950. Russia and China had supported North Korea in 1950 when it attacked South Korea, and China supported a communist coup attempt in Indonesia in

1966, which was barely averted by the U.S., with the.S. ambassador to Indonesia, Marshall Green, playing a critical role.

Ambassador Marshall Green and myself in Vietnam

When, in 1968, Nixon became President of the United States, he offered Marshall Green the job that Bill Bundy had. Marshall was a very seasoned Asian diplomat and a wonderful human being with a great sense of humor. He asked if I would stay on and work for him. I accepted with great pleasure.

Ambassador Green, right after he had been invited to take the job by Nixon, asked if I would accompany him on a trip to all the countries under his diplomatic charge and make the trip in just thirty days. In order, we went to Tokyo, Japan; Hong Kong; Jakarta, Indonesia; Singapore;

Kuala Lumpur, Malaysia; Bangkok, Thailand; Rangoon, Burma; Vientiane, Laos; Saigon, Vietnam; Manila, The Philippines; Taipei, Taiwan and Seoul, Korea. In addition, we stopped in Honolulu to meet with John McCain, the father of Senator John McCain. He was Commander-in-Chief of CINCPAC, the U.S. Central Command of all U.S. forces in the Vietnam theater from 1968 to 1972.

Wilford at a U.S. Marine military bunker in Vietnam along the demilitarized zone

During the many hours on planes going from country to country, Marshall Green and I had many thoughtful conversations about U.S. foreign policy in Asia. The topic I came back to time and time again was whether it was time to bring about a shift in our diplomatic relations with the People's Republic of China. During those conversations, Ambassador Green never told me that he might propose a change in U.S. policy toward China.

But in his trip report of April 21, 1969, that is exactly what he did. As you will note, his first recommendation on the attached first page stated: "We should nevertheless liberalize U.S. trade and travel restrictions and make it clear our historic continuing friendship for the Chinese people."

This was three years before the U.S. public was informed about the historic meeting between President Nixon and

Chinese President Mao Tse-tung to reestablish diplomatic relations between the United States and the People's Republic of China.

Below is Assistant Secretary of State Marshall Green's report, which was sent to President Nixon through Secretary of State William Rogers and National Security Advisor Henry Kissinger. Most of the report is in Appendix 1 for those readers interested in taking a deep dive into how U.S. foreign policy at the highest level is made.

The second document below summarizes the significant, top-secret steps the Nixon foreign policy team took to implement the new China policy. It is classified as "Top Secret/Sensitive—Exclusively Eyes Only," the absolute highest of secret classifications. These documents have only recently been released after fifty years under seal, and I found them recently during a trip to Nixon Library in Southern California.

6200

DEPARTMENT OF STATE

57   Washington, D.C. 20520

April 21, 1969

SECRET - LIMDIS

TO: The Secretary

THROUGH: S/S

FROM: EA - Marshall Green

SUBJECT: A View of East Asia - INFORMATION MEMORANDUM

The attached report on my trip to eleven East Asian
countries reaches a number of conclusions based in large
part on conversations with top leaders of the area and
with American officials:

1.   Mainland China: Peking has never been so extreme and
hostile as at present. This is unlikely to change as long
as Mao is in control. We should nevertheless liberalize US
trade and travel restrictions and make clear our historic
continuing friendship for the Chinese people, looking to
the post-war period. Asian leaders would understand the
political merits of such US moves.

2.   Soviet Role in East Asia: Moscow is in a deep dilemma
over how to proceed in the face of an intensifying Sino-Soviet
conflict, Indonesia's upheaval, and the costs and risks of
supporting Hanoi and Pyongyang aims. Asian leaders believe
that Moscow favors a negotiated settlement in Vietnam, but
on Hanoi's terms as far as possible.

3.   Vietnam: Our role in Vietnam (and Thailand which some
regard as East Asia's pivotal country) is widely supported.
Reports of US intentions to Vietnamize the war are welcomed
provided this is done in measured fashion. President Thieu
believes peace negotiations will reach a conclusive stage
six months hence.

SECRET - LIMDIS

MICROFILMED
BY S/S: CMS

41

In essence, Kissinger arranged with the Presidents of Pakistan and Romania to go through Pakistan on an "official four-day trip" when, in fact, he only stopped in Pakistan to disguise from the media that he was traveling to Beijing on a top-secret mission.

Here are what most of the initials stand for:

*"**Sher Ali Khan**" was head of Pakistan's mission to the United Nations;*

*"**HAK**" was U.S. National Security Advisor Henry Kissinger;*

*"**Agha Hilaly**" was the President of Pakistan;*

*"**Chou**" refers to Chou En-Lai, Premier of China;*

*"**General Vernon Walter**," the U.S. military advisor at the Paris Peace Talks regarding Vietnam;*

*"**President**" refers to U.S. President Richard Nixon;*

*"**Mao**" refers to Chinese President Mao Tse-Tung;*

*"**Ambassador Bogdan**" was the Romanian Ambassador to the United States;*

*"**Ceausescu**" refers to Romania's President, Nicolae Ceausescu;*

*"**Haig**" refers to General Alexander Haig, assistant to Henry Kissinger.*

Direct and Indirect Specific Messages Between
The U.S. and PRC

ber 10, 1969.   Sher Ali Khan tells HAK that Chinese have b
that Yahya is ready to talk about US intentions in Asia when
s Pakistan.   Sher asks HAK for more specific message for
ese, is told by HAK that Yahya might say US is removing tw
royers from Formosa Straits.   (Tab A)

·mber 19, 1969.   Hilaly tells HAK that in early November Y
told Chinese Ambassador of US interest in normalizing relat
of the two destroyers.   Peking responded that it appreciated
stan's role and had released two Americans (yachtsmen).   H
 he would give Pakistan more to say to Chinese when Chou v
e, and the President is ready to establish a secure channel.

·mber 23, 1969.   Hilaly tells HAK that further letter from Y
 that Chinese appear willing to resume Warsaw talks without
onditions and that they are worried about US-Japanese agree
Japanese militarism.   (Tab C)

·uary 11, 1970.   HAK gives text to Derksen in meeting with I
Lodge.   The message says that the President is ready to est
·re secure channel than Warsaw for matters of the most extr
itivity, either through Derksen or General Walters.   (Tab D

·uary 22, 1970.   Hilaly relays to HAK Yahya assessment of
·king about US-Chinese relations (no explanation whether this
irther contacts with Chinese).   Yahya says US initiatives hav
·uraged Chinese, who no longer see US-Soviet collusion and ·
 US should not construe Chinese readiness for meaningful dia
·n of weakness.   Chinese response likely to be measured, bu
n inclined for meaningful dialogue on all issues.   China-US v
ns very remote, with possibility of Vietnam war expansion h
ened.   HAK says Yahya should tell Chinese that we can't con
·s speculation and that we are ready for direct White House c
eking.   (Tab E)

Please review Appendix 2 if you are interested in how the most sensitive foreign policy actions are made.

I am not in any way saying that my conversations with Marshall Green, the top official in charge of U.S. policy in the region, were the critical first step in our move to reestablish diplomatic relations with China. I do believe, however, that my very early efforts during our month-long trip to Asia did help Ambassador Green and those on Nixon's staff move in that direction.

This photo was taken on Air Force One, President Nixon's official plane, on February 21, 1972, on his way to China to meet with Chinese President Mao Tse-tung and establish U.S.-China diplomatic relations.

From left: Marshall Green, Assistant Secretary of State for Asia, Henry Kissinger, U.S. National Security Advisor, William Rogers, U.S. Secretary of State, and President Nixon

As you can imagine, I learned a great deal during my years as a U.S. diplomat. I became conscious that connections can be enormously helpful in one's professional life. I realized that there was an informal network of Groton and Yale graduates who helped each other identify and promote younger Groton and Yale graduates that they thought had promise.

One of my most helpful advocates was Yale's long-time chaplain, Sidney Lovett, whom I worked with at Yale-China and who was beloved and trusted by generations of Yale Graduates, including Bill Bundy, Marshall Green, Jerry Cohen, and Bill Krebs.

It also became clear that very experienced, intelligent people can still make poor decisions, as I feel was the case with Bill Bundy about our Vietnam War strategy. With all due respect, however, I hasten to add that Bundy just might have been an (outvoted) voice of reason during top-level strategy meetings about the Vietnam War. Those were meetings with the nation's top military brass, as well as President Johnson, who, many months before, had stated that the United States, under his presidency, would never back away from his promise to win the war.

My years in Washington also gave me a better understanding of the dynamics of our political system. In the 1960s, there was a far greater willingness on the part of both political parties to reach across the aisle. This is in contrast with today's Washington, in which raw political power and money take precedence over working with colleagues of the opposite party – or other countries, to solve issues of joint concern. Senator Claiborne Pell of Rhode Island, the head of the Senate Foreign Relations Committee, is an excellent example of a Democrat who, during the 1960s and 1970s, often worked with his Republican colleagues to enact sound laws. Senator Pell was a mentor of my second spouse, Carole Angermeir, in Washington during the 70s and 80s. I found him to be a true gentleman, a characteristic that is more difficult to find in today's Washington.

Chapter Seven

FAMILY LIFE IN THE COUNTRY OUTSIDE BOSTON – GOOD TIMES

I met my first wife, Adele Merrill, early in my final year at law school in Berkeley when she was finishing up her nurse practitioner's degree at the University of California at San Francisco. We fell in love and soon married at Christ Church in her hometown of Cambridge, Massachusetts.

Before we headed to Washington for the start of my work in the U.S. Department of State, we went to Maine for our honeymoon and a bit later to Peter Island in the Caribbean. As you can (barely) see, Del caught a fish for dinner!

Del and I initially lived in an apartment at the corner of 33rd and Prospect Streets in Georgetown, which was very convenient given that I could walk to work nearby. It was also within walking distance of that part of Georgetown, where we attended various diplomatic parties and functions.

We then bought a lovely small house on Merivale Road in Chevy Chase, Maryland, just across the line from DC. It was there that we started a family when Del gave birth to daughter Ashley on December 5, 1968, two days after Del's birthday. We hung out with several good friends in Washington, particularly Peter and Wendy Benchley, Tim and Wren Wirth, Dick Holbrooke, Tony Lake and Charlie Ravenel. Del worked at Georgetown University Hospital on a research grant focused on patients who had undergone open heart surgery.

In 1972, two gentlemen, Bill Krebs and Duane Feeley, from the international consulting firm Arthur D. Little, International (ADL), asked if they could meet with me in my office at the State Department. They informed me that ADL had just signed a $3 million contract with the Asian Development Bank to undertake an "Intermodal Transportation Survey of Southeast Asia." Its purpose was to determine how public and private investments in the region's transportation infrastructure could stimulate the region's economic growth. (The unstated political rationale behind this U.S.-initiated project was to strengthen the nations of Southeast Asia in the face of communist efforts to take over Indonesia in the mid-1960s and communist North Vietnam's efforts to take over Vietnam). Krebs and Feeley pointed out that ADL had access to all the technical capabilities to take on the job but did not have any depth of knowledge of the Southeast Asian region. It was for that reason that they had come to persuade me to leave the diplomatic service and come to ADL's headquarters in Cambridge, Massachusetts, to become ADL's Director of Asian Operations and to play a major role in this assignment. They flew me to Cambridge for several meetings, and I accepted the position. One of the reasons I was ok with leaving the diplomatic service was that I had achieved the goal I had set for myself twelve years earlier in Sweden – to hopefully play a role in the establishment of diplomatic relations between the United States and the People's Republic of China.

Since ADL's headquarters were located on the Western outskirts of Cambridge, Del and I sold our house in Maryland and bought another lovely small house in Weston, eight miles West of ADL. It was there that Shandy Welch was born on August the 14th, 1970. My fondest memories were winter ones, including sledding down our steep driveway, hoping to stop before we reached the street below and not get run over by a passing car.

Another fond memory was building an igloo in the deep snow and spending the night in a sleeping bag with Hunza, our golden retriever, trying to keep each other warm and stay until sunrise. We stayed until sunrise, but I still do not understand why Hunza and I were so excited about doing this.

Once Del and I were clear that we wanted to stay in the

Ashley, Hunza, and Shandy

area, we started looking for a place to put down roots and raise our family. Like most young couples, we spread out maps and drove all over, trying to find a place nearby with land and a country feel.

We heard about an old

farm, initially built during the American Revolution by one of the Adam's family, which was just what we wanted. It was on Lincoln Road in Wayland, one town West of Weston, and called "Sheep's End." The property consisted of a worn-down, rambling house with eight fireplaces to keep the house warm during the winters, an attached barn with three stalls, a workshop, and a woodshed that could hold four cords of firewood. I had to split, saw, and fill the woodshed every year.

In addition, there was a yellow cottage in the back, a large, dilapidated, four-stall garage, and an even more dilapidated "squash court," which the previous owner had called a "squash house" in the town tax records, a place to store "squash."

The garage and the "squash house" accidentally burned down one cold winter morning. We used the insurance money to build a new garage and squash court, which you can see on the right of the painting. To the far left of the painting is the unheated sleeping porch where Del

"Sheep's End Farm" on Lincoln Road, Wayland, MA

and I slept year-round. It was ridiculously cold at times, especially when one had to get up and go into the house to use the bathroom.

Across the road was an eighty-acre field and a pond. Since the price tag was more than we could afford, I met with the three neighbors, the Beards, the Catlins, and the Farrells, to see if they would like to buy the parcels abutting their properties, with the understanding that none of us would build on any of the lands. They all said "yes" and have honored this non-binding agreement for the past fifty-five years. We could afford to buy the remaining ten acres and all the buildings for $90,000. We lived there for the next eighteen years.

Those years were very special. Del seemed to love living there, although, in retrospect, I realize that with all my traveling around the world, she carried a heavy load with two young children and a property in constant need of repair.

My constant traveling for work during the girl's formative years is one of the very few things I now regret about my life. Del was doing a great job, but a father has an important role to play during the formative years of children as well.

Del and the girls did get to travel with me on a number of occasions. For example, we all lived twice in the Philippines, once in Bolivia, a year in Taiwan, and a short time in Iran, giving them the kind of exposure to the world my father and mother had given me. Granted, Ashley and Shandy were very young.

Iran: Ashley at Persepolis

Taiwan: Shandy at a fish Market.

Above: Philippines: Del, Ashley, and a helper.
Below: The Philippines: Del, Shandy and I taking a stroll

Over the years, we vacationed in several countries, including Ireland, where we came up with the name "Ashley" while standing on a bridge over the "Ashleigh" River. Ashley, Del, and I also traveled to Tibet. From Lhasa, we drove to the base camp on Mt Everest's northern side. We then had to walk part way to Nepal due to landslides.

Shandy and Dolly

Shandy, the "country girl," had horses she loved riding on the trails around the property. On one occasion, we even helped one mare give birth to a foal in one of the stalls, a memorable experience. Shandy and I also cleared a trail on the property we named "The Nook and Cranny Trail." Almost as memorable was when the girls had two rabbits in one of the stalls, which they thought were both males, only to find one morning that the two "males" had delivered fourteen more rabbits!

We had chickens and a wonderful raccoon we named "Ranger," who we obtained rather illegally from the town "dog catcher" when Ranger was only a month old and motherless.

Of the many stories I could tell about Ranger, I vividly

Ashley and Ranger

recall one cold winter morning with snow on the ground when I let him out of his enclosure while I was using the outdoor shower a few feet away. At one point, I lifted my left arm to wash myself, and when I looked to my right,

Ranger was doing the same thing! I have particularly fond memories of both Ranger and Hunza, much like Shandy's fond memories of "Dolly."

Ashley also loved where we lived, although she had more cosmopolitan interests. Both she and Shandy attended the excellent public schools in Wayland before they went to college. Ashley went to the University of Michigan, and Shandy went to Lewis and Clark College outside Portland, Oregon, which is not very far from the farm where she and her family now live.

Shandy, Del, Wilford, Ashley, and Hunza

I look back on those many years in Wayland with great fondness. We had wonderful friends and neighbors and had numerous parties, particularly around the Yale-Harvard football game weekends when the game was played in Cambridge. Thanksgiving and Christmas were particularly special. Each Christmas Eve, the four of us would pile into a car and compete for who could see the most Christmas lights and see if we could get lost in our own town.

New Year's Day was often celebrated by twenty or more of us partying in the new "squash house." I want to honor my very close friend, Don Tucker, who I played a lot of squash with during those years and engaged in many conversations about life. Other particularly good friends during those years were Gene Tremblay, Bill Hicks, and Dan Dimancescu.

Not all of our days together were enjoyable, however, such as when Del spent many hours collecting sap from the nearby maple trees and boiling down the sap (forty units of water to one unit of maple syrup). The bad part of the story is that I accidentally knocked the knob on the stove to "high," which caused the sap to boil over and create a thick sticky mess. I recall spending the next hours cleaning up and apologizing.

Del and I invited three couples one summer for an evening excursion to a secret destination. We picked them up in our beat-up pickup truck, blindfolded them, and drove them to a nearby river. We had them take off their blindfolds and get into canoes. We told them to paddle down the river until they arrived at their destination, not telling them where that was. We picked them up thirty minutes later and took them to the pond across from "Sheep's End" farm. Once there, they were formally seated. Del and I served them cocktails, a roast beef dinner, and a fancy dessert, which we had to bring from our kitchen across the field. The evening ended with dancing in the barn.

One of my favorite pleasures on the farm was driving my 1946 Ford 9N tractor, which I named "Pepe," a name I later adopted for myself and which our grandchildren now call me. Here I am on "Pepe," lifting Del's mother, "Skipper," high up in the front-end loader bucket. She took it with a good sense of humor and showed no fear, possibly because she knew I loved and respected her greatly and would never have done anything to harm her.

Most Sundays, we attended the Unitarian church in the center of Wayland, a wonderful, meaningful experience, partly because of its minister, Ken Sawyer, who became a very good friend.

My second pleasure was cutting down trees on the property for firewood, pulling out the roots, and burning the stumps, much like I had done fifty years before at Christmas Tree Hill with our farm hand and my close companion, Carson Nile. I had to do this, however, without the benefit of dynamite.

Del's delightful father Dudley, and mother "Skipper" Merrill

For twenty years, each summer, we spent a couple of weeks at Del's family place at "Turkey Cove" near Port Clyde on the Maine coast. Across the cove, a five-minute drive away, was the summer home of Del's father's brother. When we gathered, there were about ten adults and ten children. We all got along very well and had as much fun as possible, with some mischief mixed in.

One bit of mischief remains crystal clear to this day. The Merrill clan was playing ping pong and other games up on the barn's upper floor. I gathered about six youngsters, and we went outside to where Del's cousin, Newt Merrill, had parked his car. We lifted the back of his sedan and put a log under the back chassis that caused the back wheels to spin when Newt and his family got in and tried to drive home. I then came out and fiendishly offered to drive them home in my car. However, when we arrived at their house, 17-year-old Molly Merrill grabbed the keys from my ignition and ran into the house with me in hot pursuit, but to no avail. I walked home, and we made a sign that we put in their driveway that night that read "Open House," with the desired result that they had people knocking on their door early the following day. That caused their gang to make a sign that they put in front of our house early the next morning that read "Donuts 5 cents." That trick also achieved its intended result when people driving by stopped with dollar bills in hand, hoping for some warm donuts.

Tennis, swimming, sailing, and mowing fields were some of the activities we enjoyed every summer. Del, much like her mother, "Skipper" was an excellent tennis player, and both usually won any tournament they entered. Del had a particularly powerful forearm.

Unfortunately, despite all the great times we had as a couple and family, Del and I grew apart, and after twenty-six years, we divorced. Fortunately, we remain good friends. Del lives in a lovely home she built along the shores of Turkey Cove on the coast of Maine.

Chapter Eight

ADL – THINKING SYSTEMICALLY IN A WORLD OF SPECIALISTS

The nine years I spent at ADL were among my life's most intellectually stimulating and demanding. I found myself among a group of incredibly bright, competent, and highly specialized professionals. ADL had consultants who specialized in the petroleum, food & flavors, insurance, and solar industries, to name just a few. The economics division had petroleum economists, macroeconomists, and input-output economists just to suggest a few of ADL's other capabilities. ADL had offices in London and Brussels, Japan, Saudi Arabia, and Brazil.

In this mix, I was one of the very few generalists and among the very few who knew Asia well. As a result, when consulting assignments that cut across a number of disciplines needed a leader to assemble and carry out assignments requiring many specialists, the lead management committee would often ask me to lead and assemble the needed specialized talent across the company. The following is a sampling of assignments I led or played a significant role in during my nine years at ADL.

Global scan of the world to guide Citibank's ten-year strategic plan

George Vojta, the head of Strategic Planning at Citibank, asked ADL if we could project how the world would evolve over the next ten years so that Citibank could create its ten-year corporate strategic plan based on the world that was emerging. The ADL Lead Management Committee turned to Martin Ernst, a very senior consultant at ADL, and me to lead the assignment, which lasted almost a year. I was the head of the case day by day.

We told George Vojta that while neither he nor we could predict the future, we could analyze the technological, economic, geopolitical, cultural, and other forces at work and show how they would likely influence each other and drive the world over the next ten years. This assignment gave me the tools I have used every day to analyze events taking place around the world. I developed a deep understanding of such analytical approaches as Driving Forces Analysis, Alternative Futures Analysis, and Scenario Planning. It has remained fundamental to my way of thinking to this day.

Assessing whether Toyota should produce planes for the U.S. civil aviation market

Toyota has a 100-year vision of six words: "My Car, My Plane, My Boat." Each has an engine used for personal transportation. Toyota asked if I would assemble a team, analyze the civil aviation industry, and recommend if they should get into the business of producing internal combustion-powered small planes. I brought in the former head of the FAA and a metallurgist from GE, among other team members, and we went to work. We analyzed the current industry of new planes being offered in the U.S. market and the competition from the used plane market. We then told them what they did not want to hear, that they should not enter the market. They seemed to accept our recommendation, but in fact, they kept a small staff working on a plan. Two years later, they asked me to update our analysis, which we did. I again told them that we did not recommend that they get into the civil aviation industry, and they finally accepted our advice.

The ten-year plans for the electronic industries in both Taiwan and Korea

In 1976 and 1977, the electronic industries in both Taiwan and Korea were modest, and they wanted to know how best to build these industries. I assembled a team of ADL specialists and moved with Del and the girls to Taiwan. The team spent a year working on this and submitted a report. Soon after we finished our work for Taiwan, I got a call from the Korean government asking if we would make recommendations as to how they might best build their electronics industry.

My first day in Seoul on the assignment reminded me of how tough and determined the Koreans can be. Upon my arrival at the Chosen Hotel in Seoul, I got a call from their "Blue House," akin to our White House, requesting that I come right over to meet the president's chief of staff, Dr. Yun. I was told a car was already waiting for me downstairs. When I arrived at Dr. Yun's office, he did not offer me a chair or the customary tea. Instead, he said: "In 1960, the World Bank hired Arthur D. Little to recommend how much Korea should expand our petrochemical industry, and you limited the number of refineries we should build to only two." And then he stood up for emphasis and said: "Don't you ever underestimate us again!" He then showed me out of his office. We did undertake the assignment, and the Koreans were pleased.

The "Unfinished Revolution" series of articles for the Boston Globe

On New Year's Eve of 1975, Del and I drove Tom Winship, the Editor of the Boston Globe newspaper, to a dinner party north of Boston. During the ride, I told Tom I had an idea for a series of articles in the Globe about the "American democratic experiment." I noted that April 19th of the next year would be the 200th anniversary of the Battles of Lexington and Concord that started the American War of Independence, that the Boston Globe was the leading paper in the place where the American Revolution began, and that significant birthdays and anniversaries are the times when people tend to reflect on their pasts and futures.

THE UNFINISHED REVOLUTIO

Allegiance to whom?
Can we be equal and free?
Is more better?
What dare we dream?

The cover of the first issue of the Boston Globes series, "The Unfinished Revolution"

Tom loved the idea and asked if I would run it. I said I was working full-time for ADL and could not do so, but if he hired me through ADL, I could explore these issues and produce a report on how they might best develop a series of articles for his readers to ponder. Tom hired me through ADL, and I went to work, primarily interviewing thought leaders and activists throughout the country. My friend Crocker Snow, a journalist at the Boston Globe, was then asked by Tom to create the series, which became the longest-running series of articles in Boston Globe history.

Demonstrate to the Saudis that Mobil Oil was a good corporate citizen.

I was asked to go to Saudi Arabia and give a short course to Saudi business leaders on international business. At the time, Mobil Oil was about to be nationalized by the Saudi government, and Mobil wanted to demonstrate, a bit belatedly, that they were good corporate citizens. When I first went to Saudi Arabia to meet with Mobil Oil's chairman, Frank Jungers, he spoke to me forcefully when he said, "If you blow this assignment, ADL will never work for Mobil Oil/Aramco again!" This was a real threat and not a comfortable one, given that my assignment was in the tens of thousands of dollars while ADL's annual revenue from Mobil/Aramco was quite a few million dollars. I wrote five short case studies on international business and had them translated into Arabic. I then flew back to Saudi Arabia and delivered five lectures to about 150 Saudi business leaders from the country's Eastern region. Many of the attendees arrived at the Al Gosaibi Hotel in Al Khobar by camel. Chairman Jungers was pleased, and I was relieved.

Designing the City of Yanbu in Saudi Arabia

Concerned that Iran or Iraq might attack Saudi Arabia to

gain control of the Saudi oil fields and refineries in the East of the country, the Saudis hired ADL to design an oil refining city in Yanbu in the West of the country along the Red Sea 1,000 miles across the country. I was asked to analyze how they might best get laborers from around the region to do the work to build the city, given that no self-respecting Saudi citizen would do manual labor. I analyzed the possible sources of manual laborers from the region and concluded that Egypt would be the primary source. When I first visited Yanbu, there were only 500 inhabitants, mostly fishermen and their families. Now, thirty years later, there are 335,000 inhabitants.

Railroad project from Baghdad to the Turkish Border

The head of ADL's office in Brazil contacted the main office in Cambridge to say that the largest construction company in Brazil, Constructor Mendes Junior, was bidding on a project in Iraq to build a railway from Baghdad to the Turkish border. They wanted to hire ADL to think through how they would manage the project. The ADL lead management committee had no idea who they should ask to take on the assignment until one member said: "Wilford H. Welch knows the Middle East fairly well and loves taking on assignments he knows nothing about. He would relish the opportunity to figure it out." So, off to Baghdad, I went.

NASA and the commercialization of space

The 1970s were still early days for NASA. They had put a man on the moon at great cost and were being pressured to come up with ways the private sector might be interested in paying NASA to do research in space. I was asked to assemble a team and develop recommended areas for space research in a near-zero gravity environment.

Teaching International Business Management

When I arrived at ADL, I found that they had a small management education program primarily for government officials from developing countries in the Middle East and Africa, which was paid for by the U.S. Agency for International Development. (USAID). I was intrigued and explored whether they were teaching courses in international business management. They were not. I found no textbooks on international business management at the Harvard Business School or any other business school in the U.S.

I volunteered to create and teach a course entitled "International Business Management." I started teaching a nine-session course of 90 minutes each and expanded it to twenty sessions after several years. Given that there was no textbook on the subject, I had to create case studies based in large measure on the international consulting assignments I had been involved in.

After several years, I had 70 students in my class from all over the world. And as I had learned in Hong Kong a number of years earlier, I again found that I loved to teach. Many days, I felt that I was just a few minutes ahead of my students. I taught this course for eight years and loved every minute of it. One year, I had five students from Iran. They invited me to go to Iran to teach my course there and have one of them continue to teach my course at their business school after I left the country. I gave the Iranian student I selected all my materials for him to use.

In the Spring of 1976. Del, Ashley, Shandy, and I flew to Iran, and I taught them while they traveled around the country. We all enjoyed Iran immensely due to its rich history and lovely people. Iran's citizens would soon be radicalized by the revolution in 1979 that continues to this day.

Toward my latter years at ADL, the company decided to invest in me by paying for me to attend the thirteen-week Program for Management Development (PMD) at the Harvard Business School. It proved to be a valuable learning experience. By then, I had written an article for the Harvard Business Review entitled "The Business Outlook for Southeast Asia," drawing heavily on the work we had done for the Asian Development Bank a few years before (see my article on the next page).

Harvard
Business Review

The magazine of decision makers

May-June 1973

As I have suggested, my years at ADL were very demanding, educational, and stimulating, just as I had hoped. My learnings from those years were many; I learned the critical methodologies for better understanding the forces changing our world during our assignment at Citibank, to think strategically in systems, and as a futurist, I confirmed how much I enjoyed teaching; I was exposed to many more countries and cultures, particularly in the Middle East; I developed confidence that I could take on almost any challenge.

Chapter Nine

THE WORLDPAPER

"THE VOICES OF THE WORLD SPEAKING ON THEIR OWN BEHALF ON ISSUES OF GLOBAL CONCERN"

In the Spring of 1980, Crocker Snow called to tell me that he had left the Boston Globe to become Editor-in-Chief of a new international publication called "The WorldPaper." He asked if I would consider leaving ADL and becoming the Publisher of the parent company, The World Times. The more I looked into this, the more intrigued I became, primarily because of its unique editorial mission, "The voices of the world speaking on their own behalf on issues of global concern." In those days, most international reporting about the world came from Western journalists writing from a Western point of view. After nearly a decade with ADL, I accepted and left for the WorldPaper. The WorldPaper's editorial approach was totally different in another way. At its editorial core were fourteen well-respected journalists from fourteen regions of the world. For example, Silviu Brucan, our associate editor for Eastern Europe, was a key member of the communist takeover of Romania in 1947 and, for several years, was the Romanian Ambassador to the U.S. and the UN. He then turned on Romanian President Nicolae Ceausescu, who in turn put Silviu under house arrest twice. Silviu eventually got even when he sat in judgment of Ceausescu and his wife in a rump trial on December 25, 1989. Ceausescu and his wife had tried to flee the country by helicopter but were forced down near the Romanian border. A rump trial was organized, and both he and his wife were found guilty, taken outside, and shot by firing squad.

Our other Associate Editors were equally distinguished.

A Chinese gentleman learning about the world from the WorldPaper

WorldPaper Associate Editor Muctar Lupis had been imprisoned for years in Indonesia for criticizing the government. Wang Zongyin, our Associate Editor from China, had been on "The Long March" with Mao in 1934 -1935; Yoshiko Sakarai was not only a journalist but a well-known TV personality in Japan; Amitabha Chowdhury from India was based in Hong Kong, Hilary Ng'Weno was from Kenya. Our European Associate Editor, Jacquelin Grapin, was from France and also wrote for Le Monde; Osama El-Sherif was from Aman, Jordan, and Alexander Pumplanski was from Russia.

Every year, we would bring all the Associate Editors together somewhere in the world to discuss the forces at work and story ideas. Some of our major investors came to these gatherings as well, which they saw as a benefit of their involvement with the WorldPaper. On occasion, we invited our advertisers as well, including Kim Armstrong, the head of all advertising at AT&T, and Marshall Carter, the CEO of State Street Bank.

The WorldPaper's editorial approach to topics was also unique. Many issues would include different perspectives on a single topic. For example, one of our cover stories was on the changing role of women. The WorldPaper provided three perspectives from three women native to the countries they were writing from: one from the U.S., another from China, and a third from Saudi Arabia.

Regarding the WorldPaper's distribution system, every month, the English edition had to be translated into the five other languages used by the twenty-seven newspapers around the world that carried the

WorldPaper as an insert into their newspapers. That called on us to produce films sized to the presses of each paper in each county and send them to them by DHL, the international carrier service.

Our distribution was largely in the developing countries of Asia, Latin America, and the Middle East. To get leading papers in the U.S. and Western Europe to carry the WorldPaper was a tougher sell because they already had their own journalists stationed around the world.

We were, in essence, asking each publication's editor to give up editorial control over serious WorldPaper editorial content with points of view they might not agree with. That was a particularly tough sell to the editors of well-established publications, such as the People's Republic of China.

I had many different roles as we built the WorldPaper. On some trips abroad, I met with the editors-in-chief of publications that we wanted in our network. If the editor-in-chief was open to the idea, he (or she) and I would then meet with the publisher to explore his or her willingness to pay for its publication. I met with many U.S. companies and their ad agencies to sell advertising space in all our editions or in select regions of the world. I was the head of two major research projects, the first being the Wealth of Nations Index, in which we analyzed the economic, social, and technological capacities of twenty-eight developing countries to achieve balanced economic, technological, and social growth. (For a summary of our research, please see Chapter 17).

Another research project, The Information Society Index, assessed the readiness of fifty-five countries to take advantage of the Internet revolution. Occasionally, I would write an article about a topic I had considerable knowledge about.

In 1994, I left the WorldPaper to move to San Francisco. It had been a fascinating decade during which I learned a great deal more about the world. I became even more aware of the need to listen to the perspectives of other cultures that we can all learn from. That, in turn, enabled me to reassess some of my unconscious assumptions about my own country, particularly our "economic growth at all costs" approach that I will write about later in this book. For example, while I continue to believe that capitalism is far superior to other systems when it comes to creating economic growth, I began to question the fact that it gives too much emphasis to growth and profits "at all costs," which I refer to the longer-term consequences of our exploiting the earth's natural resources and people, particularly indigenous peoples.

2000

IDC/World Times
Information Society Index

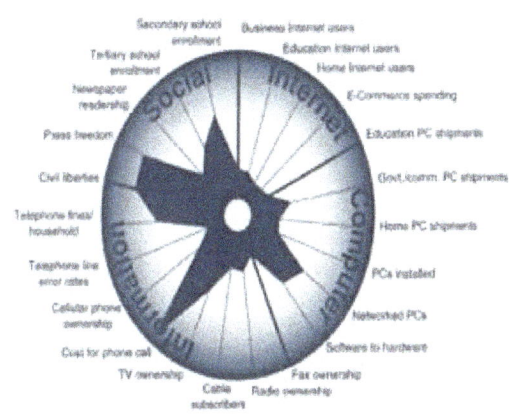

Analyze the Future

http://www.idc.com
http://www.worldpaper.com

*Measuring the Global Impact of Information Technology
and Internet Adoption*

Chapter Ten

ENGAGING WITH NGOS I FELT WERE DOING IMPORTANT WORK RELATING TO MY INTERESTS

Over the years, I became deeply involved in several not-for-profit organizations whose missions resonated with me. There were four in particular: The National Outdoor Leadership School (NOLS), NatureBridge, The Tibet-U.S. Resettlement Project, and Columbia University's School of International and Public Affairs (SIPA).

The National Outdoor Leadership School (NOLS)

NOLS has been the world leader in teaching wilderness skills and risk management for over sixty years. I had not heard of NOLS until 1980 when I was organizing a trip for my wife, Del, Jamie McLane, a Yale Classmate and later the Chairman of Outward Bound, and me to undertake a thirty-day trek around the Annapurna range in Nepal.

I felt I needed to develop more skills to deal with any possible serious mishaps during our trek. I considered taking an advanced first aid course but finally signed up Del, Jamie McLane, and myself for a month-long NOLS course in the mountains of Wyoming, where NOLS is headquartered. That started my love affair and involvement with NOLS for the past forty-five years.

Over the years, I have also taken a NOLS sea kayaking course in Alaska and a mountaineering course during which we summited Mount Denali, formerly known as Mt. McKinley, during a twenty-nine-day climb. Our leader was the renowned NOLS mountaineering instructor, Steve Goryl, who has since become a close friend. Steve Barr, a relative and wonderful friend of mine, joined me, which made our already close bond even closer.

Steve Barr and me on a sunny day at 15.000'

At 20,310 feet, Denali is the highest mountain in North America and Europe and is very dangerous. With NOLS, however, I felt quite safe.

On day 20, we climbed fifteen hours from 14,000' to 17,000' through very deep, heavy snow. Five Japanese climbers descended along our route at one point and were constantly falling. They were delirious, frightened, and exhausted. They had summited three days before but had then been hit by a tremendous snow and windstorm that took their leader's life and caused all the other climbing team members to get frostbite, which was the reason for their falling. To save themselves, they left their dead comrade at 18,000'. Our NOLS leaders, Steve Goryl and Jim Chisholm, made sure we were okay at our camp, and, despite having just climbed fifteen hours, they climbed up and brought the Japanese leader's body down. This photo, which I took the next day, shows a rescue team lowering his body down a very steep ravine.

Digging three feet down in the snow to protect our
tent from treacherous winds

Lowering the body of a Japanese climber down from 17,000 feet from camp 3 on Mount Denali

During the climb, I recorded our adventure daily. NPR interviewed me upon my return and produced this eight-minute segment that was played on NPR's Morning Edition. You can find it at WilfordWelch.com.

After our successful ascent, my godson, Barclay Welch, an experienced climber, expressed interest in also climbing Denali, but ideally from the northern side, an even more challenging route. While Barclay had never taken a NOLS course, John Gans, the head of NOLS while I was chair of the board, questioned whether Barclay was up for NOLS' most difficult and dangerous course. I told John that I could vouch for Barclay's competence both physically and mentally, and that was all John needed. Barclay's father, Harry, my oldest brother, however, had a very different message for his "little" brother. Harry said to me: "If Barclay dies on this NOLS trip, I do not know if I could ever forgive you." I was sure Barclay would be safe given his skills as a climber, his good judgment, and the fact that NOLS is so safety conscious. Barclay had the climb of his life.

In 1994, I organized a team of five NOLS trustees to provide financial support for five NOLS climbing instructors who planned to summit Mount Everest from the Nepalese side. The expedition was called the "Sagarmatha Environmental Expedition," and "Sagamartha" was the Nepalese name for Mount Everest.

My godson Barclay Welch on his Denali climb

The expedition's goal was not only to summit the 29,032-foot mountain but also to remove 5,000 pounds of climbing trash from the high camps. This goal is in line with the NOLS motto of "Leave No Trace," or stated otherwise, "If you packed it in, you pack it out."

The climb achieved both its objectives and Scott Fisher and Rob Hess became the second and third U.S. climbers to reach the summit of Everest without using oxygen.

My niece and goddaughter, Hilary Welch, joined me on this expedition. In the following photo, she is wearing a white jacket. She held her own among all the men.

The NOLS climbing team and several trustees in support of the Everest expedition

The trek to base camp was strenuous, and once at base camp, some of us climbed into the very dangerous Khumbu Ice Fall, which has massive crevasses and vertical walls of ice that can only be climbed by putting up ropes and ladders, as shown in the next two photos.

Climbers crossing the Kumbu icefall. Photo by Scott Fischer.

Steve Goryl climbing nine ladders roped together. Photo by Scott Fischer

Steve Goryl at camp four on Mount Everest at 26,000'

This photo is a selfie taken by Steve Goryl at Camp Four on Mount Everest at 26,000 feet. Steve stayed there by himself for three nights, waiting for "good summit weather." When Scott Fischer and Rob Hess came down to Camp Four after reaching the summit, Scott said to Steve, "Come down with us, or you will die." Steve declined the offer, saying that he would take his chances. He ate out of frozen tins left by previous climbers and used oxygen from the many oxygen bottles also left by climbers. The weather cleared, and he summited on his own.

Scott Fischer died on Everest two years later when he was the leader of one of the two teams made famous by the book "Into Thin Air," during which eight climbers died, including another group leader, Rob Hall of New Zealand.

This was my second visit to Mt. Everest, the first being in 1980 when Del, our daughter Ashley, and I reached the Everest Base Camp on the Tibetan side of the mountain. It was on that trip that I took this photo of Mt. Everest from the Tibetan side.

National Geographic cover photo of their climbing map of Mount Everest
using my photo taken from Tibet

Almost by pure chance, it became the National
Geographic Society's cover photo of the Mt. Everest
Climbing map produced by Brad Washburn. He and his
team had spent years producing an extensive map of the
mountain that noted all the routes that had been

climbed, each of the deaths on each route, and their cause.

Brad asked me over dinner at our farm in Wayland "if I had taken a photo of Mount Everest from the Tibetan side in the fall of any year when some of the winter snow had melted, thus showing more of the rock face?" Then and there, I went up into our attic and gave him the Kodachrome photo that was used on the map's cover.

In 2023, I gave a presentation to NOLS field instructors on how global warming and climate change are likely to impact the environments around the world on NOLS' courses in mountaineering, rock climbing, sea kayaking, backcountry skiing, and coastal sailing. You can watch the presentation on WilfordWelch.com.

Video presentation to NOLS instructors on how climate change will impact courses

Nature Bridge

When we moved to the San Francisco Bay area in 1994, I learned of another outdoor education school that intrigued me. At that time, it was called the "Yosemite Institutes," but it changed its name to "NatureBridge" because there were six campuses in the United States, and the name "Yosemite" was confusing. NatureBridge provides experiential outdoor education to children aged 8 – 16 in each of its campuses.

By that time in my life, the notion of educating our citizens about the importance of the natural world had become very clear. After a year or so on the NatureBridge Golden Gate board of trustees, I was elected chairman and remained chair for the next eight years. The Golden Gate campus is located in the Marin Headlands, four miles from our home in Sausalito. With few exceptions, most of our students are sponsored by their schools. Our courses are 3-5 days, and the students stay on campus in our dorms. It was because of my connection to NatureBridge that I developed very close friendships with Cleve Justis, who at that time was the head of the Golden Gate campus; Also, my wonderful publisher friend Raoul Goff, who I nominated to the national NatureBridge board; Ben Toland and Dave Jones, who I nominated to the NatureBridge Golden Gate board, two good friends who continue to take on significant wilderness adventures, and who have also been deeply involved in NOLS. Their spouses, Deborah Jones and Laurie Durnell, are also good friends of both Carole and mine.

The Tibet-U.S. Resettlement Project

While still back on the East Coast in the late 1980s, I learned about an effort to enable 1,000 Tibetan refugees living in camps along the Indian-Tibet border to come to the United States for permanent resettlement. I joined the board and soon became its co-chair with the Dalai Lama's representative in the United States, Rinchen

Dharlo. We were able to entice Massachusetts Senator Ted Kennedy to get the U.S. Congress to pass a law to allow this resettlement project to move forward. We set up a system in which the Tibetan authorities would make sure that a cross-section of Tibetan men and women would be considered and that they would not choose only well-educated professionals, such as doctors, and thus drain the remaining refugees of badly needed services.

Carole already had met His Holiness the Dalai Lama a number of times both in India and in Washington, D.C., and she was able to arrange for us to receive an invitation for the two of us to meet with His Holiness in Dharamsala, in Northern India. Dharamsala is where he had lived in exile ever since 1952 when he fled over the Himalayan mountains as the new communist regime invaded Tibet.

When we met the Dalai Lama in Dharamsala in 1994, I told him that I only had one question, "How can we create a program that assures that after two or three generations in the U.S., the Tibetans will not have lost their Tibetan culture? His response was immediate. He said that would not be possible because, unlike the Jews, who have been forced to move all over the world for centuries and have learned how to take their culture with them, Tibetans have had no such experience or skills.

With His Holiness the Dalai Lama in India in 1994.
Image by Carole Angermeir

To address that issue, the Tibet-U.S. Resettlement Project team decided to establish twenty communities in the U.S. where fifty Tibetan refugees would be relocated within a fifty square mile radius of each other so that they could maintain relations and continue to practice their traditional ways. We put the word out, and in a very short time, we were able to find twenty towns in the U.S. and fifty families in each who were willing to have a Tibetan refugee live with them for a year, provide them with health care, and help them find a job. I was blown away by the generosity of so many of my fellow Americans, something I will never forget when I feel we are moving away from empathy for others and towards self-centeredness.

Columbia University's School of International and Public Affairs (SIPA)

My good friend, Michael Hoffman, from my ADL years, was chairman of the board of SIPA. He invited me to join the board, where I served for six years. SIPA educates over 700 graduate students from all over the world each year in an excellent world affairs program. Those among the ten board members were Brent Scowcroft, President Bush's national security advisor, and David Dinkins, the former mayor of New York. When at Yale, I developed the impression that Columbia was not a great institution, largely out of ignorance and my seeing them through the lens of their football teams, which were poor. During my years at SIPA, I found Columbia to be a most impressive university.

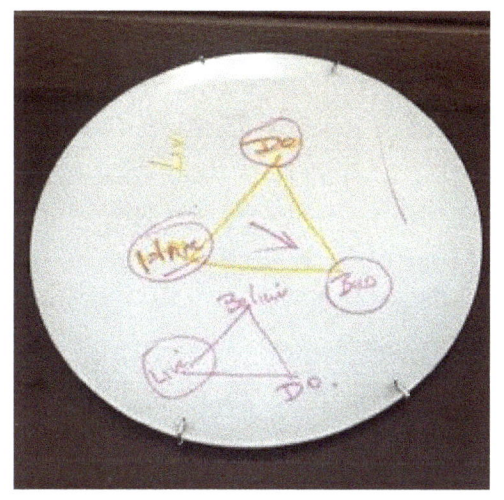

Chapter Eleven

THE ART OF THE SWEEPAWAY – KEEPING LOVE ALIVE

In the early 1990s, I was in New York doing business and was invited to the 28th birthday celebration of a good friend from Boston, Geralyn White. Before dinner, I was introduced to Carole Angermeir, the hostess. When we met, I was struck by her beauty, but as our conversation continued, I sensed that her greatest beauty came from the inside. She was thoughtful and curious; she had a great sense of humor and was surprisingly vulnerable about some of the challenges she had faced in her life. I seem to have had enough experience, accumulated over my 52 years, that I decided to pay attention.

Sometime during the meal, we were all asked to explore a question and draw what came up on a plastic plate. I suggested to Carole that we explore whether she was drawn to a life defined by "Doing," "Having," or "Being."

I drew a triangle, on which each of us marked where we were in our lives and the direction we wanted to head, and we then discussed our answers. The plate with our results is shown here, in which we both noted that we wanted more "being" in our lives.

It was some time before we connected again, but one day, I got the news that a close Yale friend and Berzelius secret society member, Dave Karetsky, had just been killed in an avalanche while skiing in deep snow in the Bugaboos in Canada. Because his Jewish faith called for burial within three days, the service was to take place at a synagogue in New York in three days, even though his body was still in the Bugaboos being examined by the local coroner. It was important to me that I honor my dear friend by going down to New York for the service. On a whim, I called Carole to see if she was available for tea before the service. She accepted, and it was then that we found ourselves attracted to each other.

Sometime later in our evolving relationship, Carole called with a fun idea: "Could you meet me at the airport next Friday for a 4 PM flight? You will be back on Monday afternoon." When I asked where we were going, she responded, "I am not going to tell you." When I asked what I would need to bring, she said, "Your passport and a pair of skis."

Given that it was not ski season, and the closest skiing would be in Patagonia, Chile, where I could rent skis, I concluded that she was playing with me and only a passport might be needed. Thus, our first "Sweepaway" was launched.

When we arrived at Orlando Airport in Florida, I feared that she was taking me to Disney World. When we went to the Delta lounge, and she implied that we were waiting for a car that would drive us to our destination, my worst fears were amplified.

When we were leaving the Delta lounge, Carole asked me to close my eyes. She put a blindfold on me and a sign around my neck. When she led me to the elevator and we went down to the terminal, I knew we were either going to a car or a plane. When people walked by me, a blindfolded man with a sign around his neck, they shouted things such as "Ah, India!" or "Ah, Paris!."

And while I was thoroughly confused, I was also enjoying the moment. I soon could tell that we were getting on a plane, where fellow passengers, as we walked down the aisle, said things like, "Ah, South Africa." And, when we were buckled into our seats, with me still masked, the captain came on the intercom and said, "Welcome aboard. Today, we are going to a secret destination"! It was very clear that Carole Angermeir knew exactly how to conduct a secret "Sweepaway"!

Twenty minutes into the flight, Carole let me take the blindfold off, and I got to work. My first clue was that we were on an Eastern Airlines flight. My second clue was that it was a Boeing 757 that flew relatively short routes, and that would exclude a flight to Europe and certainly to Patagonia. My next clue was that the location of the sun indicated that we were flying South.

As we were approaching our destination, the stewardesses all came over to us as a group, and the captain again came on the intercom and proclaimed, "Welcome to the Bahamas!" We had a wonderful time over the next two days.

Thus started our tradition of inviting the other on "Sweepaways," which might be for a weekend or more in length. "Whiskaways" are for a day or overnight, and

"Slipaways" tend to last for only a few hours. The rules are that the person inviting must choose an experience that they know the other would really enjoy and an experience that would be safe. Here is the original "Sweepaway" sign Carole put on me thirty years ago that I recently used when I took her on an overnight "Whiskaway." We find that the gifts we give each other from time to time are not only great fun but also keep our love fresh and alive. We hope to write a book someday on this subject, entitled "The Art of the Sweepaway – Keeping Love Alive."

Taking Carole on a "Sweepaway." She has no idea where I am taking her

We have reflected on what has enabled our love to grow

and deepen over the years. It seems it is the result of the values we share in common, including deep trust and respect for the integrity of the other, a willingness to be vulnerable, and a "love 'em anyway" attitude to each other's quirks, a notion we learned from our friend Richard Russell, and his song with the same name.

In the mid-1990s, we moved to San Francisco to start a new life. We were married in 1997 and have been happily married for over twenty-eight years. Below is a more recent photo at a lively wedding.

Carole and me dancing at the country western themed wedding of Michael Hoffman and Vanessa Farnell in the Cotswolds

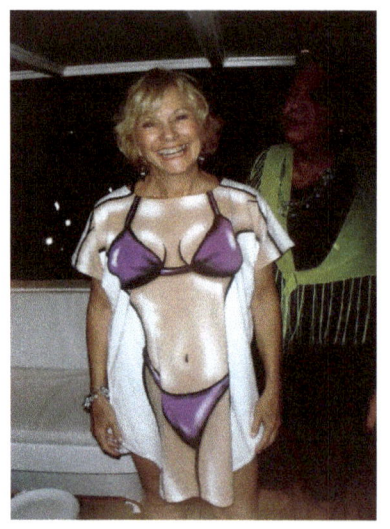

Here is Carole in a tee shirt having a good time on a boat with friends off the coast of Turkey during our honeymoon.

For the past twenty-five years, we have lived on a houseboat in Sausalito, California, on the eastern end of the Golden Gate Bridge. I love it primarily because we have a sense of community on our dock of forty houseboats at a time in our country when we seem to be moving from "relationships" to "transactions."

We love the diversity of professions and backgrounds of our fellow houseboat neighbors, and I love being so close to nature, which is highlighted twice each day by the ever-changing 5–6 foot tides. These are daily reminders of who is ultimately in control of our destiny. Stated otherwise, Mother Nature, not the human race, will show us "who bats last." A photo of our houseboat is below. If you would like to see more photos, please put in your browser "VRBO 79742."

During the summer months, we have boat parties, as you can surmise from the photo below. Great music, great wine from the Napa Valley, swimming, and laughter.

"Sitting on a Dock of the Bay" by Otis Redding was written here in 1967

Chapter Twelve

CROSS-CULTURAL JOURNEYS – EXPLORING THE WISDOM OF OTHER CULTURES

While living in New York in 1991, Carole, at the request of her good friend, Helen La Kelly Hunt, joined a small group of women to form the founding Board of Trustees of the New York Women's Foundation. The purpose of the NYWF was to fund and support projects by women who were devoting their time and life energy to run grassroots projects serving the many unmet needs of women and children in the boroughs of New York City. Carole had observed that many of the women philanthropists she was working with at the NYWF could both benefit from and enjoy travelling with groups of women to countries where they would be challenged and empowered.

This observation, and her year-long research project in Mexico studying indigenous ways, was the genesis of *Women-to-Women Cross Cultural Adventures.* She created trips to expose women to unique destinations and cultures where they would feel that they were doing something both challenging and rewarding. One of her first trips was to sleep in tents in very rural Tanzania and interact with tribal women. Other travel companies at the time had itineraries to see the tourist sights. CCJ was a different way to travel.

Carole has a background in anthropology and had done fieldwork in the Yucatan region of Mexico for a year in 1970, in Playa del Carmen, where she and her two young children lived on the beach in a Mayan village for a year. They built and lived in a "casita tipica" with a thatched roof.

CCJ brochure with many journeys offered

The "casita tipica" on the shore of Playa del Carmen, Mexico that Carole and her children built and lived in for a year

There were only twenty Mayan families living in Playa del Carmen at the time, and no outsiders besides the three of them. There were no roads, no phones, and no one had ever left the village. Her children, Teresa and Charlie, ages 8 and 10, attended the local school.

Carole drove the first person to leave the village to a dentist three hours away to have all his teeth pulled without anesthesia. When Carole returned to pick him up, he was sucking on a lollipop to make him feel a bit more comfortable. I recently read her report about her year-long field trip in that isolated village, which I found fascinating. I think the insights she gained during her field research made it possible for her to conceive of establishing trips that would focus on exploring the wisdom and ways of indigenous peoples.

After 3 years, she found the women-only trips limiting, so she changed the name to Cross Cultural Journeys and opened CCJ's journeys to both women and men. This proved to be a better business model.

For the next twenty-five years, Carole built Cross Cultural Journeys to become a substantial travel company. By the time she sold CCJ in 2015, she was offering trips with well-known international leaders to over thirty countries, including Tibet, Ethiopia, Cuba, South Africa, Botswana, Morocco, Egypt, Bali, Australia, Nepal, Mongolia, and Iran.

Carole treating a San Bushman in rural Botswana

I went on many of the CCJ trips and led a few. I developed a group of trips called "Countries at the Crossroads," which included South Africa, a trip I led with former U.S. Ambassador to South Africa, Princeton Lyman; a trip to Cuba co-led by former U.S. Consul General to Cuba, Wayne Smith; and a trip to Northern Ireland, co-led with well-known Belfast journalist, Frank Costello. One of the members of our group to Northern Ireland that we scholarshipped was Aquella Sherrils. Aquilla was the former leader of the Crips gang in Los

Angeles who had fostered a peace treaty with the opposing gang, the Bloods. Both sides of the Northern Ireland conflict were excited to hang with and learn from Aquella.

Two Hamar tribal women in Ethiopia

CCJ's first trip to Ethiopia was to the remote south to spend time with the Hamar tribe, an area where no other travel company, including the National Geographic Society, had ever taken travelers. After floating across several streams by Land Rover, as there were no roads or bridges, her group stumbled upon a coming-of-age ceremony. A young man had to successfully run over the backs of forty lined-up cattle and back again without falling. If he fell, he would have to leave the village in disgrace. If he was successful, he could stay and choose a bride. Here is the cattle jumping scene she and her group happened to come across.

On another of Carole's early women-to-women Cross Cultural Journeys trips to Southern Ethiopia, Carole and her group of ten travelers came across a "tea house" being built by eight Ethiopian women who were widows. These women, the CCJ women travelers, were told they could not own or work land on their own. Only men could till the land. The only problem with that custom was that most of the men in the area had been killed during Ethiopia's recent war with Eritrea.

Carole with the blind village woman in Ethiopia, who was the leader of the Tea House project

The widows had built a Tea House to earn money for themselves and their children, but Carole learned that the women did not have the funds to put on the roof. She huddled with the ten American women traveling with her, and each agreed to put in $500. She then asked the women how much they needed to have the thatched roof built. They huddled and came back with the answer of 500 birr, which is the equivalent of $16 U.S. dollars! The remaining $4,984 was the seed money Carole used to establish the Cross-Cultural Journeys Foundation, which, 28 years later, is still active. The purpose of the foundation is to provide travelers on CCJ trips the opportunity to fund projects in the countries they visit. Carole and I visited the leader of the tea house project in

Ethiopia many years later. She was totally blind, yet instantly recognized Carole's voice. The tea house was being continued by the daughters of the founders.

I learned a great deal more about the world and myself during these CCJ trips. In particular, it deepened my awareness and capacity, as we said at the WorldPaper, "to listen to the voices of the world." In the case of CCJ, the learnings were largely the voices of indigenous peoples.

Carole with children on a CCJ trip to Mongolia

Another initiative was two major conferences in Bali, entitled "Quest for Global Healing." Three of us created and ran this initiative, namely Carole, Marsha Jaffe, another Sausalito resident, and me. I was responsible for the conference speakers and content. My primary objective was to shine a light on the pain and suffering that most of the world's population faces every day, of which Americans are often unaware. We chose Ubud, Bali, as the location and the grounds of the ARMA Museum and Resort, owned by a wonderful Balinese friend, Agung Rai. I knew the Nobel Peace Prize recipient, The Reverend Desmond Tutu of South Africa, well enough to ask him to be our marquee personality, and he accepted.

The first Quest for Global Healing gathering took place in 2003, was attended by 250 guests from twenty-five countries, and lasted five days. It was so successful that we organized another one in 2005 at the same location, attracting another 450 people from 40 countries. What seemed to make it so special was that it was culturally rich, exposed attendees to people and ideas from many countries around the world, and was just plain fun.

One example of our speakers awakening attendees to the pain in the world was Arn Chorn-Pond of Cambodia, who shared his experience in an open-air prison in the mid-1970s. At that time, the communist-led Khmer Rouge slaughtered up to two million of the country's 7.5 million citizens, mostly intellectuals. Arn had been told that he would be kept alive so long as he played his flute to help make the prisoners more peaceful while they were being bludgeoned to death with baseball bats. I asked Arn how many people he witnessed being killed. He first said he had no idea. I asked if it might have been 100, to which he responded, "much more." He then reflected and said, "Probably close to 5,000."

It was in Ubud that we met for the first time our dear friend Nadya, an extraordinary designer of beautiful, exotic clothes, and another very special friend, Zan Zan, and his wife, Putu.

Arn Chorn-Pond of Cambodia who stayed alive in Khmer Rouge death camps by playing the flute

Below: Desmond Tutu dancing at the conference

Below: Desmond Tutu again dancing at the conference!

Below: Carole, Desmond Tutu, Wilford, and my daughter Shandy Welch

Me with Desmond Tutu, and two other Nobel Peace Prize recipients; Jody Williams of the U.S. for her work eliminating land mines, and Betty Williams of Northern Ireland for her work on religious wars

Balinese dancers at the Quest for Global Healing Gathering

Balinese dancers at the Quest for Global Healing Gathering cont'd

Chapter Thirteen

EXPLORING THE WORLD'S SUSTAINABILITY
AND GLOBAL CLIMATE CRISES

In 2010, I started focusing intensely on the global sustainability crisis. I define this crisis as the world having a population that is more than the earth can support. This is in large measure due to people in the more advanced economies exploiting and consuming the world's natural resources at a far faster rate than they can be replenished. There are myriad examples, from the exploitation of fresh water to agricultural land, the destruction of vast tropical rain forests, the overfishing of all edible species, and the polluting of our rivers and oceans.

I started to focus on the ways that social entrepreneurs were creating initiatives to provide health, education, transportation, microcredit, and other services to the poor in developing countries. In addition, I became interested in initiatives of social entrepreneurs who are addressing human rights and social justice abuses and providing disaster relief to those people in countries who will be the hardest hit due to the global warming/climate crisis.

This led to a wonderful relationship with a Middlebury College junior, David Hopkins, who appeared in my office in Sausalito one day and declared that he was going to work for me whether I paid him or not. As soon as David graduated, he joined me. Over the next year, he and I met with, or interviewed over the phone, 100 or more social entrepreneurs all over the world who had created successful social enterprises. (And yes, I paid him).

Social entrepreneurs create not-for-profits that are doing good in their fields while earning the funds to reinvest in their enterprises and earn a modest living for themselves. We were primarily seeking to understand the business models they developed and encourage others to go into the field. This resulted in the book *The Tactics of Hope, How Social Entrepreneurs are Changing our World*. Over the next year, David and I spoke as often as possible at world affairs forums, business schools, libraries, and elsewhere. David and I have remained very close friends ever since.

In 2010, I organized a gathering on the big island of Hawaii called "Beyond Sustainability – Creating a Culture of Leadership on a Platform of Reverence." Its purpose was to invite indigenous leaders, particularly from the Hawaiian Islands, to share their beliefs and practices that the rest of the world might adopt to address the global sustainability challenge.

In 2015, I concluded that while sustainability was a critical challenge for humanity, there was an even greater challenge facing all of us: the global warming/climate crisis. Humans, for over a hundred and fifty years, have increasingly been releasing CO2 into the atmosphere at rates that are warming up the planet to such a degree that we are facing great peril if we do not reduce our consumption habits and our use of fossil fuels. The fires in greater Los Angeles during much of January 2025 are just one of hundreds of examples. The fires destroyed 13,000 homes and 300 commercial buildings, including churches, schools, and hospitals, for an estimated loss of $30 billion.

Rows of houses and businesses turned to rubble by the wildfires in Las Angeles

I made the shift because it had become clear to me that global warming could quite literally result in the world's "6th mass extinction." This could wipe out the entire human population, along with most plants and animals, unless we quickly take some very uncomfortable steps, individually and collectively. And, while the world is making some progress, I am growing less and less confident that we are mature enough as a species to address global warming aggressively and decisively and in time to avert disaster. The choice of the future we will have on this planet is entirely up to the human race.

As stated otherwise, I am concerned that while the human race is extraordinary in developing new technologies, we are in our adolescence in terms of our ability to deal with the consequences of our technological advances and the complexities of our societies. I see the culprit being our focus on short-term gains to satisfy our near-term wants rather than thinking longer term and taking actions within our power now to protect the futures of our children and grandchildren. For example, polls by Pew and Gallop showed that during the 2024 U.S. Presidential election, "climate change" ranked between 15th and 18th in importance.

Our inability to put a damper on our near-term wants and focus on the actions we need to take today to address our fossil fuel emissions just may kill us. The metaphor of "moving the chairs on the Titanic" as the steamship sinking comes to mind.

It is for these reasons that I wrote a book entitled *In Our Hands – A Handbook for Intergenerational Actions to Solve the Climate Crisis,* which I have been updating every few years.

It is tragic that in the United States, the global warming/climate crisis has been politicized, which has serious implications for our ability as a nation to focus on solutions. Our politicians should not argue about whether global warming is real, but instead, they should focus on the most effective ways to address it for the sake of all those who come after us.

By this time in my life, I feel the values I hold that drive the actions I take are well entrenched and will not change over the rest of my life. These years did, however, reaffirm my love of teaching and my laser focus on the global warming/climate crisis.

I also believe in the power of hope and the surprising ability of Americans to turn the tide when pressed to the wall - as was so eloquently expressed by Winston Churchill: "You can always count on the Americans to do the right thing - after they have exhausted all the other possibilities."

The challenge is heightened by irreversible "tipping points," defined as warming events in parts of the world such as Greenland, Siberia, and Antarctica that we cannot stop once they reach a certain temperature. We have ten years at the most before some of these "tipping points" are likely to be reached. The two graphs below summarize the realities I am convinced we are facing.

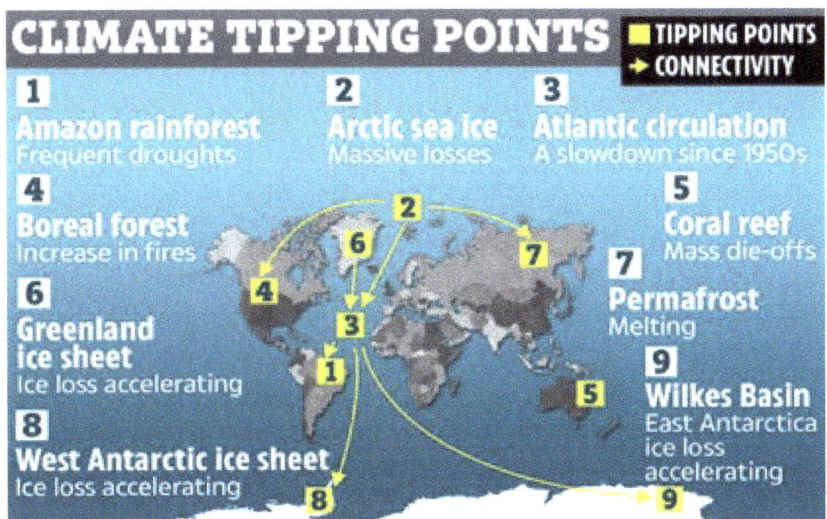

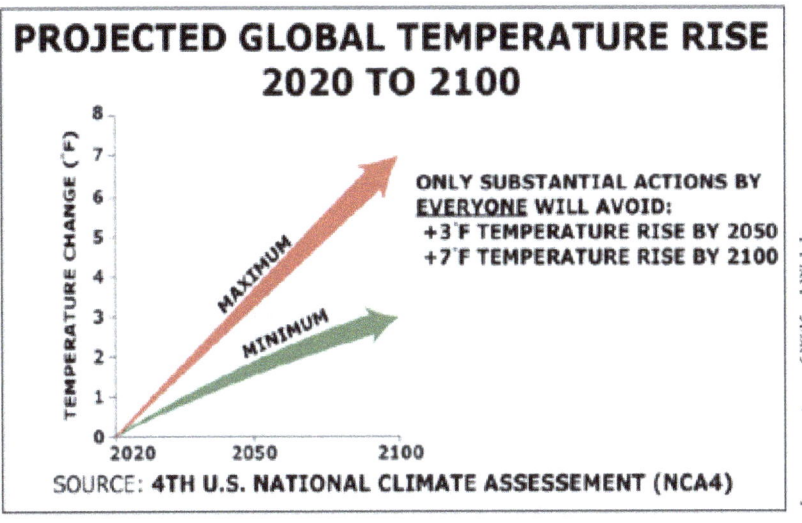

I wanted the book to be used by teachers and their

students in K-12 classrooms throughout the United States, given that global warming/climate change will very likely seriously impact the lives of all K-12 students today during their lifetimes and the fact that the subject is not taught in most schools. With this in mind, I concluded that the Presidio Graduate School in San Francisco would be a good place to start.

I went to the dean, the provost, and the lead professor teaching sustainability and got polite expressions of interest - but no action. I thought I had failed to get their attention until a year and a half later when I got a call from Liz Maw, the president of the Presidio Graduate School, who said that she had concluded that I was right.

We agreed to work together to make it happen. Over the next few months, a team was assembled to develop a virtual educational program for K-12 teachers throughout the United States that teachers could draw upon in their various classes. I became the lead teacher, and "In Our Hands" became the primary text. The results have been impressive. Presidio marketed the program directly to K-12 teachers throughout the country and to the heads of school districts. For the last few years, I have been sitting at my computer teaching a six-week course entitled "Climate Essentials" to twenty or more teachers from all states in the country. We then created a four-week course entitled "Climate Justice" and another four-week course entitled "Extreme Weather." I feel like "a kid in a candy store," for I was back teaching and was teaching the subject that will clearly be my professional focus for the rest of my life.

I am currently writing another book on solutions to the climate crisis, entitled "Your Choice."

Chapter Fourteen

FAMILY

Fortunately, the two daughters Del and I have, Ashley and Shandy, are thriving.

Ashley and Shandy

Ashley lives outside of Boston, near where she grew up. She has built a successful consulting firm, Summersault Innovation, that trains the sales forces of corporations such as GE, Microsoft, and SalesForce to turn sellers into customer-centric problem solvers. As I type this, she is in Beijing to deliver one of her programs there.

Ashley and John McGah's daughter, Noelle, graduated from the University of Wisconsin two years ago, has been working for Planned Parenthood in Chicago, assisting those women who can legally get abortions, and is now getting a nurse practitioner's degree. Aydan has just graduated from the University of Michigan and is moving to New York to start a computer programming job.

Ashley, WW and Noelle at her graduation, Aydan and their father, John McGah

Shandy and her husband, Hans, and their two daughters, Finley and Ayla, live on a beautiful farm twenty miles south of Portland, Oregon.

For two decades, Shandy dedicated her career to improving patient care and developing clinical orthopedic programs for a large healthcare system. She then launched her own leadership coaching business, where she partners with individuals and teams to discover their potential, enhance communication skills, and master the art of leadership. She also writes a weekly blog entitled: "Weekly Wisdom," which is excellent. You may want to subscribe.

Ayla Moller, Shandy Welch, Finley Moller, and Hans Moller

Finley and Ayla Moller aren't as attached to farm life as their parents but enjoy spending time with the animals as well as their human friends. Finley dreams of becoming a singer, while Ayla is focused on sharpening her soccer skills and backpacking.

Shandy with a friend

Meanwhile, Hans, an orthopedic surgeon, continues his passion for advancing the healthcare system. He recently expanded his community involvement by joining the local school board. On their farm, they have one llama, two horses, two cows, five sheep, two dogs, two cats, and a pet raccoon who comes and goes.

And with another friend

I believe I have been a good father and a good friend to a number of people around the world, but I am also clear that all the "doing" that I have touched on in the preceding pages has meant that I was not as available to Ashley and Shandy as they were growing up, as I now wish I had been.

I have now given myself permission to step back from so much doing and focus more of my energy on Carole's and my family members. That "Doing, Being, Having" triangle that Carole and I discussed over the birthday dinner the night we met, during which I drew an arrow from "doing" to "being," is rather belatedly taking place.

In 2015, Carole sold Cross Cultural Journeys after twenty-five years and is now in a well-deserved retirement. My former wife, Del, lives in a spectacularly beautiful area along the coast of Maine. She and I chat on the phone at least once a month and have wonderful discussions about life, our children and grandchildren, and the state of the world.

Finley, Aydan, Ashley, WW, Shandy, Ayla, and Noelle

Interesting how genetics plays out. I was one of four boys; Del and I had two daughters, and they, in turn, had two daughters. This photo shows how very fortunate I am.

Of my three brothers, my oldest brother Harry died recently at 95, my brother Perry died four years ago, and my sister Carolyn died in 2010.

At our wedding on August 24, 1997, at Old St. Hilary's church in Tiburon, California

Minister Paedar Dalton, Perry, Wilford, Harry, Noble, Carolyn

I want to write a bit more about Carolyn, whom I was incredibly close to, possibly because she always started her letters to me with the playful and positive phrase "Dear Wild and Wonderful Willy." I am very proud of Carolyn, for she also explored paths that were not taken throughout her life. Right out of graduate school in Washington, D.C., she heard of an organization being

Carolyn in the field with NPR

created called "National Public Radio." She went to see the only person in a basement office, and within minutes, she became NPR's first employee and NPR's "Director of Research." She worked there for 38 years before her passing in 2010.

For many years, Carolyn produced audio stories called "Radio Expeditions," which were eight-minute segments on NPR's "Morning Edition." Radio Expeditions programs also appeared on the radio program of the National Geographic Society. I remember vividly one segment when she and her husband, NPR Morning Edition host Alex Chadwick, were deep in the jungle of the Republic of Congo. First, you heard an elephant let out a kind of desperate scream. Then you heard why – the sound of chainsaws and the falling of giant trees.

One of Carole's many gifts to me over the years was to organize a Welch family gathering at a fishing camp in Oquossoc, Maine, to celebrate my 80th birthday and brother Harry's 90th. Even more special than the birthdays was the fact that virtually all Welches from around the United States and several very close friends showed up for three days and nights of friendship and fun.

Welch family reunion at a fishing camp in Maine in 2018

In the summer of 2023, Carole and I created a special trip to Tuscany for several of her and my family members.

From left: Aydan McGah, Wilford, Carole, Sophia Johnson, Noelle McGah, Teresa Faatz, Ashley Welch and Charlie Warner in Tuscany

Carole had two children from a former marriage, Teresa Faatz, age 63, and a son, Charlie Warner, age 62. Tragedy struck both in the spring of 2024 when Teresa died of cancer in March and Charlie died of cancer in June. The grief Carole feels is almost unbearable, and it is very challenging for me as well.

Carole with her brother Stan, his wife Anne, Charlie, her father Paul, and granddaughter Kennedy

I would like to say something about Carole's father, Paul Angermeir. I came to respect and love Paul very much. He did not know that he had dyslexia until Carole went to a conference about dyslexia when Paul was 65 and suggested that he be tested. This changed his life, and he read voraciously for the rest of his days. Paul was a gentle, beautiful soul who lived in gratitude and was very close to Carole and her brother Stan. He died gently two months shy of his 100th birthday. All of us deeply miss him.

Charlie and I became particularly close during his last two years while he was trying to stop his cancer from spreading, particularly during the six months he moved in with us.

The photo below was taken when he and I rode from San Francisco to Los Angeles a few years ago in support of AIDs. We biked 600 miles in six days. It was a surprisingly easy ride, in part because it was all flat.

Charlie and me at the start of our 600-mile ride and horsing around

Charlie leaves behind three daughters, Ashley King, Kennedy Warner, and Sophia Johnson, a granddaughter, Coralee, and a grandson, Jackson. He also leaves behind his dear wife, Misty 'Stanton' Warner, whom he married eleven days before he died. To the right of the wedding photo below is Misty's delightful daughter, Astrid.

From left: Daughters Kennedy Warner and Sophia Johnson, Charlie, Misty Stanton Warner and Misty's daughter Astrid

Charlie and Misty getting married in our living room on May 30, 2024. Our good friend and neighbor, Elana Yonah Rosen, who married them, is in the foreground.

This photo is of Charlie's eldest daughter, Ashley King (in the far back of the car), on the trip to Cuba that Charlie organized as his dying wish for his daughters and grandchildren, his mother, and me. On the front is her husband Joe, and on the back is granddaughter Coralie and grandson Jackson. Also, one of Charlie's other daughters, Sophia Johnson.

Good Friends

At Groton School, and over the many years since then, I developed life-long friendships with many classmates and teammates, including Sam Webb, Gordon Gund, Hugh Scott, Bill Polk, Ken McLean, and Stewart Forbes, among others. Here is a photo of some of them and their wives during a trip to Cuba which Carole organized.

From left: Stephanie Mashek, Hugh Scott, Donna Scott, Stewart Forbes, Deede Forbes, Ken Maclean, Bill Polk, Lulie Gund, me, LuAnne Polk, Gordon Gund, and Carole

Top row from left: Terry Thomas and David Tise. Middle: Bill Twist Bottom: Vince Ricci, WW and Charles Hobson

Another group I get together with every month is a men's movie group that has been going for over forty years, although I have been a member a mere twenty.

About twenty-five years ago, Allen Macomber, a fellow NOLS trustee, and his wife Geraldine invited Carole and me, and a number of their other good friends, to join them in the south of France to celebrate Allen's 60th birthday. We all stayed at the Chateau des Aspras in Correns. We loved it so much that we had friends join us for a week the next summer. Since then, Lynne and Bill Twist have made a tradition of going to Chateau des Aspras. This photo, taken during one such trip, will give you a sense of the fun and friendship we enjoyed together.

Not in order are Bill and Lynne Twist, Ian and Victoria Watson, Gordon and Lilly Starr, Charles and Sandra Hobson, Terry Thomas, Vince Ricci, Dianne Morrison and Michael Hebert, John van Merkensteijn. Carole and me

And here is another escapade with many of the same friends while on a private boat amongst the Greek isles

Fifteen years ago, eight of us created a group we called "The Great Hearts," the best-known being Dr. Jerry Jampolsky, who is particularly well known for his first book, "Love is Letting Go of Fear." Jerry founded The Center for Attitudinal Healing, which he and his wife, Dr. Diane Cirincione, led for many years. We meet on the first Wednesday night of each month for a simple dinner to enjoy each other's company and to share how we are doing in our lives. We do not give each other advice but listen deeply. We are witnessing each other's journey.

The Great Hearts - Judy Kimmel, Wilford, Carole, Jerry Jampolsky, Diane Cirincione, and Joel Kimmel. Missing are Bonny Meyer, Mark Finser, and Don Goeway

The dock where we live has forty houseboats. Over the 24 years Carole and I have lived here, a real sense of community of people who genuinely care for each other has emerged. For example, two years ago, when an 80-year-old neighbor's houseboat sank during a storm, within an hour, a neighbor offered her a place to live for the next six months. In a country that, during my lifetime, has moved increasingly from relationships to transactions and from concern for country and others to concern mostly for self, this is a very positive place to live one's life.

We have many friends in the houseboat community, and I will not seek to mention them all here. In addition to Joe Novitski, who I have mentioned earlier, and his partner Susan Huxtable, we see a great deal of Gordon Haight and his partner, Linda Meyer, Elana Yonah, Bruce Thomas, Ted Bravos and his partner Joanie Kendall, Don and Rebecca Lytle, Bonny Meyer and Diane Cirincione.

East Pier residents at one of our "Wine Down's" from 5-6 PM each Friday during the summer months at the end of our dock

Other close friends in the Bay area that I would like to mention were Galen and Barbara Rowell, who died ten years ago in a plane crash, returning from giving lectures on a ship in Alaska. Galen was one of the world's greatest outdoor photographers, and Barbara helped greatly in his business. They are sorely missed. Also, Jonathan and Frannie Field have been so generous to me over the years, as has the delightful Kathy Evans, a great poet. I also want to mention Randy McNamara, who was the lead trainer at EST and the Landmark Forum for years and now is a close friend and collaborator with me on solutions to the climate crisis.

During Carole's twenty-five years running Cross Cultural Journeys, Margarita Ramirez was critical on a day-to-day basis, as was Sybil Sanders, CCJ's accountant. Margarita's husband, an irrepressible Irish minister, Paedar Dalton, married us.

During Carole's twenty-five or more CCJ trips to Cuba over the years, one of our guides was a wonderful Cuban named Walkys Acosta. Walkys had expressed interest in moving with his family to the United States. I went to work with former U.S. State Department colleagues and ultimately helped get them visas to immigrate. They sold their apartment in Havana, told almost no one about what they were planning to do, took all the funds they had, and got on a plane to Mexico. They then rode a bus to the U.S. border, where they were let into the U.S. The next day, the U.S. government announced a change in immigration policy, which would have left Walky, his wife, Yadira, and their daughter, Rocio, stranded in Mexico. They now live in Florida, where Walkys teaches English as a second language and Yadira practices as an accountant. They remain good friends of ours.

In preparation for one of CCJ's trips to South Africa, we met Susan Collins Marks of South Africa in Washington, DC. She was lovely, well-connected and well-informed.

Carole and I were drawn to her immediately and have developed a close relationship with Susan and her husband, John Marks. John was a pioneer in the field of both social entrepreneurship and conflict resolution. He founded Search for Common Ground in 1982, which remains the world's largest non-profit focused on conflict resolution in the world's most dangerous countries. Carole had known John for decades before Susan also came into our lives.

I mentioned earlier my close friend and Yale classmate Jamie McLane, but I have not mentioned "Lucy," the imaginary cat his son, Josh, and I have shared for the past forty years. One Sunday morning, when Josh was about five, I cradled my arms and asked Josh if he would like to pet the kitten I was holding. He could not see it until I asked him to look closer and again asked Josh if he wanted to pet "Lucy." He then could "see" that Lucy was there and started holding and petting "Lucy" himself. Josh and I, to this day, still ask each other how "Lucy" is doing.

And finally, our very best friends, Ian and Victoria Watson, from London and Cortes Island, which is off Vancouver Island, British Columbia. Ian and Victoria joined us on our honeymoon, cruising off the coast of Turkey shortly before our marriage. Our wedding reception was at their home, on the lagoon about two hundred yards from ours in Belvedere, California.

When it was time for Carole and me to leave for our honeymoon, Carole and I walked out to their pier. Carole threw the flowers to the bridesmaids, and we turned around, took off our kimonos, jumped into the lagoon, and swam to our home and the start of our second honeymoon.

We have stayed with Victoria and Ian at their summer home on Cortes Island in British Columbia for the past sixteen summers. One of our most meaningful trips together was to sail and visit the Haida indigenous peoples on the islands of Haida Gwaii, 500 miles north. Our captain was our good friend Albert Zeman.

Wilford, Carole, Victoria, and Ian Watson

For the past five years, Carole and I have spent an hour on Zoom each Saturday with Victoria and Ian checking in with each other. Their daughter, Lucinda, is our wonderful goddaughter. Other close friends on Cortes are Charles Steinberg and Torkin Wakefield, Rick Ingrasci and Peggy Taylor, Rupert Sheldrake and Jill Purce, Robert Gass and Judith Ansara, John Blaxall and Deepa Nayaran.

Our goddaughter Lucinda Watson with Carole

Carole speaking her truth at a rally in Sausalito in March of 2025

Chungliang Huang, also known as "Al" Huang, is also a good friend of ours and of Ian and Victoria.

For years after his escape from China, as it was being taken over by the Communists, Al became the leading Tai Chi Master in the United States and is beloved by many thousands of Americans and others from around the world. Given my background in China, it was not surprising that we soon became good friends. In 2004, Chungliang and I led a CCJ trip to China, which, because of his knowledge and contacts in China and his effusive personality, proved to be a very special journey for all. He was, of course, the principal leader of the trip to his homeland.

Chapter Fifteen

OBSERVATIONS ABOUT AGING

There appear to be somewhat predictable phases in one's life, such as Phase 1: Being taken care of and educated; Phase 2: Earning money for one's family and retirement; and Phase 3: Being retired and spending more time with family and friends.

I now see myself belatedly focused less on all the "doing" that is reflected in these pages and more on "being," by which I mean seeking to be fully present with others and with myself each moment. (That, at least, is my goal). While writing this book, I have come to the realization that during most people's professional lives, as a banker, a diplomat, an attorney, a teacher, a builder, or other work, we tend to highlight those characteristics that we feel are appropriate to our profession.

Throughout most of my professional life, I have sought to demonstrate to others that I am worthy of their trust and respect, particularly regarding the work I might be doing with them. But at this point in life, as I am approaching ninety, I am no longer seeking to impress others so that they will want my services. I feel no need to seek anyone's approval or be anything except who I am. This shift has enabled me to explore in some depth who the authentic me is and how I want to live the rest of my life.

A perfect example of this transformation was Charlie Warner, who prided himself for forty-five years on being "a Green Builder," just as I, at times, have presented myself as a diplomat, a teacher, a publisher, etc. Two years ago, when cancer forced Charlie to stop being a "Green builder," a beautiful, gentle soul emerged. He loved and was good as a "Green Builder," but what emerged was the real "Charlie Warner." Charlie's beautiful transformation taught me a valuable lesson.

When I first heard Alan Jackson singing "The Older I Get," I felt he was singing about the values I now seek to live by. These are the lyrics:

The older I get

The more I think

You only get a minute; better live while you're in it

'Cause it's gone in a blink

And the older I get

The truer it is

It's the people you love, not the money and stuff

That makes you rich

And if they found a fountain of youth

I wouldn't drink a drop, and that's the truth

Funny how it feels. I'm just getting to my best years yet

The older I get

The fewer friends I have

But you don't need a lot when the ones that you got

Have always got your back

And, the older I get

The better I am

At knowing when to give

And when to just not give a damn.

And if they found a fountain of youth

I wouldn't drink a drop, and that's the truth

Funny how it feels. I'm just getting to my best years yet

The older I get

And I don't mind all the lines

From all the times I've laughed and cried

Souvenirs and little signs of the life I've lived

The older I get

The longer I pray

I don't know why; I guess that I've

Got more to say

And the older I get

The more thankful I feel

For the life I've had and all the life I'm living still.

The lyrics of this song reflect one of my favorite aphorisms, "I hope to die young, but as late as possible."

If I had a chance to do my life over again, I don't think I

would change a thing, even the pain I have occasionally had to deal with, both physically and emotionally. Actually, as I have written previously, I wish I had spent more time with my daughters during their childhood than I do now with them, my grandchildren, and Carole's family members.

I am very clear that of the 8 billion people on our small planet, I have been among the most fortunate. For starters, I was born in the United States. Secondly, I was born into a loving family who had the commitment and resources to enable me to get a good education. Thirdly, I missed the First and Second World Wars, during which more than 50 million people were killed or injured. Fourthly, I missed the Great Depression.

All this is to say that by chance, I happened to live at a time of great economic growth that enabled me to do all the traveling and work I have been privileged to do around the world.

I also have to acknowledge that I have had a lot of help along the way, first from my family and second from connections with some very influential men. It is crystal clear that a number of well-established, prominent men gave me excellent opportunities along the way. The group includes Jack Crocker, the headmaster of Groton, Sydney Lovett, the longtime chaplain of Yale, Bill Bundy and Marshall Green of the U.S. Department of State, and Bill Krebs of ADL.

That is the type of support columnist David Brooks seemed to be writing about in the December 2024 edition of The Atlantic entitled "How the Ivy League Broke America." He makes a good point that there were a number of people in Ivy League universities during my time who got into those universities because of connections. They may have gotten good jobs because of their father or mother's connections, whether they were qualified or not.

I'd like to believe that I got the jobs I had mentioned in these pages from the men mentioned above because I was qualified, and those men (and they were all men at that time when women had not yet broken through the "glass ceiling"), were all looking for what they believed to be "up and comers." If I had failed to live up to the expectations of any of them, I would never have gotten another call. I was definitely also in a privileged position, in part because of the schools I went to, the contacts I made there, and what I achieved in those schools.

Chapter Sixteen

OBSERVATIONS ABOUT THE EVOLUTION OF MY VALUES

As I reflect on what I have written regarding the phases of my life, I realize that the following values have been enduring:

- Coming from love, not fear (Footnote 1).
- Curiosity/Exploration of the world and ideas.
- Living each day with gratitude.
- Determination.
- Integrity.
- Being a team player.
- A systems approach to understanding the world.
- A deep commitment to the natural world.
- Empathy for the plight of others.
- A desire to be of service.
- Seeking a balance between doing, having, and being.

Footnote 1) Several years ago, when Carole and I were having dinner with our good friends Greg and Mary Thomson, also of Sausalito, Greg asked me what I thought were two of the most important values or attributes of a healthy, meaningful life. I responded almost immediately with the words: "Love and curiosity." A week later, Greg sent me this song he wrote and played on his guitar entitled "Love and Curiosity." You can find this on wilfordwelch.com

- A desire to cause others to feel "seen" and acknowledged in a country in which much of the population has moved to "transactions" rather than "relationships." Once or twice a day, I will ask a cashier or some other service person: "How are you doing?." Hopefully, they will feel that I am asking them a serious question and that I want to acknowledge them. It almost always puts a smile on the face of the other and causes me to smile as well. If you resonate with this thought, you might like to read New York Times and Atlantic columnist David Brooks' recent book "How to Know a Person."

Don Quixote on my desk

- A positive attitude and a belief that almost anything is possible. I have this wooden embodiment of Don Quixote on my desk primarily because Don Quixote "dreamed impossible dreams," and I believe that just about everything is possible.

Many people limit themselves, while others seek to make what most people believe impossible possible. We harnessed electricity; We put a man on the moon. We are dreaming about going to Mars. We developed AI, actions that most humans could not have imagined.

While writing this book, I have also come to realize that

the values I hold have indeed been fundamental to the decisions I have made. Even the meditation I say every morning is filled with values. It goes like this:

> *"I will live today with deep gratitude, with love, with integrity, with intention, with graciousness, with gentleness, with awe and with joy. I will seek to be totally present, aware, curious, coming from cause, not effect, centered, and authentic in all that I say, do, and feel. And I will treat you, my body, with respect, in the exercise I give you, the nourishment I give you, and the rest I give you."*

Chapter Seventeen

OBSERVATIONS ABOUT FORCES THAT ARE CHANGING OUR WORLD

I feel I am in a position to provide an educated assessment of where our nation and our world are heading. I do not have a crystal ball, but I have explored most of these topics in some depth during my lifetime.

I feel that this is also true because of the use I have made of the driving forces analysis and alternative futures methodologies I learned during the Citibank assignment at ADL and have used them ever since.

These assessments have also had the benefit of my education and exposure to the law, business, economics, and history, and what I have learned about our world while living, working, or traveling to well over 100 countries.

There are many positive developments made by the human race, such as the fact that modern medicine has enabled the average human's life expectancy to double since 1900. In addition, incredible medical advances, particularly in pharmaceuticals and medical practices, have occurred and will continue, most notably breakthroughs driven by artificial intelligence.

However, I am going to end this book by focusing first on a number of trends that concern me, I am not seeking to frighten you, but to focus your attention on the forces I see at work that are likely to do great harm to you, your children and your grandchildren. They are all challenges that you, and all of us, can do something about now that will make their futures better. Every challenge I mention in the pages that follow were created by humans and can be solved if we focus and take individual and collective action now.

The world's population is unsustainable

Our global population of 8 billion is not sustainable, given how we allocate the world's rapidly diminishing resources. This growth in the world's population and the lack of resources to go around, coupled with the global warming/climate crisis, is already decimating many people around the world, such as those in Somalia and the Sudan. Lack of fresh water will be the major killer, in part due to fights over it. At the end of this book, there is a very short chapter from my previous book, "In Our Hands," which presents a worst-case scenario of what humanity could be facing unless we take better care of the natural world, which we depend upon for our survival.

The world's extractive "economic growth at all costs" model is unsustainable and is the primary cause of our global warming/climate crisis.

There is no question that capitalism, as it is practiced in the U.S. and in many other countries, will result in more significant growth in GDP than in countries run by communist and socialist regimes. Yes, China has had far faster rates of economic growth recently than the U.S., and that has been the result of well-thought-through five and ten-year plans and massive central government initiatives and subsidies. China's impressive growth was also the result of Chinese Premier Deng Xiaoping's declaration in the early 1980s, "To make money is beautiful," which made it communist government policy for Chinese entrepreneurship to flourish, a cultural and economic capacity that Russia has never had.

The capitalistic model, however, focuses on "economic growth at all costs," which often results in both the diminishing of the planet's resources and what's left of indigenous peoples living traditional ways. And, instead of seeking to stay within nature's "limits to growth," we continue to destroy critical ecosystems and focus on new ways to extract resources, such as the increasing focus on opening seafloor mining of minerals like lithium, which are in short supply.

ECONOMIC ENVIRONMENT

NATIONAL ECONOMY
- GDP growth rate
- Real GDP per capita (in PPP$)
- Annual inflation rate
- Savings rate
- Government deficit/surplus (% of GNP)
- Current account balance (% of GDP)
- External debt/GDP
- Debt service/exports
- Foreign reserves (excl. gold)/imports

INTERNATIONALIZATION OF THE ECONOMY
- International trade (% of GDP)
- Speed of integration
- Convertible currency?
- Foreign direct investment
- Portfolio investment
- Market capitalization

BUSINESS ENVIRONMENT
- Economic Freedom Index
- Independent Central Bank?
- Full central clearing services?
- Commercial Property Protection
- Privatization Index
- Corruption Perceptions Index

INFORMATION EXCHANGE

INFORMATION APTITUDE
- Newspaper readership (per 1000 inhabitants)
- Literacy rate
- College students studying applied and natural sciences (%)
- College students studying abroad (%)
- English as primary language for business?

INFORMATION INFRASTRUCTURE
- Radio ownership (per 1000 inhabitants)
- Television ownership (per 1000 inhabitants)
- Telephone ownership (per 1000 inhabitants)
- PCs (residential) in use (per 1000 inhabitants)
- Faxes in use (per 1000 inhabitants)
- Cellular telephones in use (per 1000 inhabitants)
- Membership in INTELSAT/ITU/WIPO?
- Information technology expenditure (% of GDP)

INFORMATION DISTRIBUTION
- Books published annually (per 100,000 inhabitants)
- Press Freedom Index
- Number of independent daily newspapers published
- Number of independent radio stations
- Number of independent television stations
- Cable television available?
- Satellite television coverage?
- Number of Internet service providers

SOCIAL ENVIRONMENT

STABILITY AND DEVELOPMENT
- Income distribution
- Male/female wage parity
- Unemployment rate
- Refugees as % of population
- Territorial disputes?
- Political rights index
- Independent Rule of Law
- Private automobiles/trucks in use (per 1000 inhabitants)

HEALTH
- Birth rate
- Life expectancy
- Death rate
- National healthcare program?
- Government expenditures on health (% of GDP)
- Pension spending (% of GDP)
- Population per physician
- Grain harvested area per capita
- Daily calorie supply

NATURAL ENVIRONMENT
- Protected land
- Air pollution index
- Population with access to clean water
- Signatory to CITES/ITTA/Montreal Protocol/UNCLOS

One of the seldom discussed reasons we seem trapped in a growth-at-all-cost system is our national debt, which now requires yearly interest payments greater than our military budget. We are on a train that has no off-ramp. In addition, our political system causes both parties to add to the national debt by legislating more spending. As a result, our system may crash if we cannot pay the interest on our national debt. As stated otherwise, we are trapped in a trap of our own making with potentially dire consequences.

I favor approaches that seek to "redefine prosperity" that would move us away from only pursuing economic growth at all costs to one that values such things as being in nature and enjoying "being" as well as "doing" and spending.

In 1989, while working for the WorldPaper, a colleague and I developed "The Wealth of Nations Index." (See below). We used 63 variables, each given equal weighting, to measure 35 emerging countries. The variables are mentioned below. Note that one of the variables we measured was called "social environment," which included "natural environment" measures as well as "health" measures.

And here were the resulting rankings:

MARCH 1997 RANK	COUNTRY	CHANGE SINCE SEPT. 1996		ECONOMIC ENVIRONMENT		INFORMATION EXCHANGE		SOCIAL ENVIRONMENT		TOTAL SCORE
				Score	(Rank)	Score	(Rank)	Score	(Rank)	
1	Taiwan	▲	1	544	(1)	462	(3)	391	(7)	1397
2	South Korea	▼	1	540	(3)	517	(1)	332	12	1389
3	Chile		0	470	(5)	438	(4)	397	5	1305
4	Czech Republic		0	386	(7)	416	(6)	418	2	1220
4	Israel	▲	1	384	(8)	465	(2)	371	10	1220
6	Malaysia		0	541	(2)	352	(13)	296	21	1189
7	Argentina	▲	1	382	(10)	437	(5)	332	12	1151
8	Hungary	▼	1	298	(25)	408	(7)	401	4	1107
9	Poland		0	305	(21)	363	(12)	430	1	1098
10	Mexico	▲	1	353	(11)	374	(10)	305	(17)	1032
11	Thailand	▼	1	482	(4)	266	(21)	272	(27)	1020
12	Costa Rica	▼	1	305	(21)	321	(16)	392	(6)	1018
13	Uruguay		0	241	(32)	392	(8)	378	(9)	1011
14	Brazil		0	351	(12)	343	(14)	308	(16)	1002
15	Venezuela	▲	2	294	(26)	391	(9)	279	(25)	964
16	Russia		0	268	(30)	372	(11)	304	(18)	944
17	Romania	▼	2	250	(31)	380	(20)	411	(3)	941
18	Colombia	▲	1	304	(24)	315	(17)	302	(20)	921
19	China	▲	5	399	(6)	223	(27)	277	(26)	899
20	Panama	▼	2	286	(27)	282	(19)	320	(15)	888
21	Ukraine	▼	1	204	(34)	323	(15)	355	(11)	882
22	Indonesia		0	383	(9)	206	(28)	284	(24)	873
23	South Africa		0	342	(14)	305	(18)	224	(33)	871
24	Turkey	▼	3	346	(13)	227	(24)	296	(21)	869
25	Jordan	▲	2	319	(17)	227	(24)	291	(23)	837
26	Tunisia		0	317	(18)	191	(29)	327	(14)	835
27	Ecuador	▼	2	272	(29)	254	(22)	304	(18)	830
28	Philippines		0	331	(16)	225	(26)	259	(28)	815
29	Peru		0	309	(20)	241	(23)	234	(30)	784
30	Egypt	▲	1	314	(19)	178	(32)	255	(29)	747
31	Morocco	▼	1	305	(21)	189	(30)	230	(31)	724
32	India	▲	1	332	(15)	152	(34)	215	(35)	699
33	Pakistan	▲	1	284	(28)	189	(30)	224	(33)	697
34	Cuba	▼	2	150	(35)	156	(33)	387	(8)	693
35	Vietnam		0	238	(33)	126	(35)	228	(32)	592

©WORLD TIMES GLOBAL RESEARCH/THE WORLDPAPER

The future of the human race will likely become unsustainable if we do not get off fossil fuels far faster than the human race seems capable of doing.

No matter what President Trump proclaims, the global warming/Climate crisis is real. It was driven by mankind during the Industrial Revolution and gained speed as we ramped up our collective desire to do and enjoy more things, such as our wanting more combustion-driven cars using petroleum and flying all over the world using aviation fuel. All such actions have decimated the land due to mineral exploitation and the destruction of the world's forests, which are critical to taking fossil fuel gasses out of the atmosphere.

As a result, we are now living in the "Anthropocene era," in which man has more influence over the natural world than she has always had over us. We decimate the earth in all sorts of ways, which suggests we are in control, but the fact is that the CO2 we are putting into our atmosphere, as all 8 billion of us strive for more, will result in Mother Nature ultimately having the final say. We may warm our atmosphere and earth to such a degree that the planet becomes uninhabitable.

I am an optimist by nature but a realist about the climate crisis. I have devoted much of the past two decades to understanding the science and probable impacts of global warming/climate change, as well as solutions to solve it. I have increasing concerns, however, about the human race's capacity to address a crisis that does not, to most people, seem immediate, and will interfere with they desire to have more and more, now.

Nearly all members of the human family are short-term oriented, not unlike the hunter-gatherers who had to look over their shoulders to survive. I believe we are in our adolescent phase regarding our capacity to address longer-term challenges, such as global warming/climate change. We live our lives much like little boys, wanting all the toys they can get and wanting them now.

Unfortunately, it has become clear to me that the IPCC (The Intergovernmental Panel on Climate Change) is not up to making hard, enforceable decisions to reduce fossil fuel emissions based on science. All 195 countries in the IPCC have to agree on the wording of any official IPCC communique, and thus, the conclusions and actions are watered down. Saudi Arabia and the petroleum and natural gas industries during the second Trump administration have been able to hijack and block all efforts to get the world to slow down its use of fossil fuels fast enough to avert disaster. And just as concerning is the fact that the commitments made by the member states are not enforceable.

A recent article in the Wall Street Journal captures my current thinking: "Climate optimism is fading. Higher costs, pushback from businesses and consumerism, and the slow rollout of technology are delaying the transition from fossil fuels. It will become increasingly impossible to face the changing climate we will experience."

During the first Trump presidency, I engaged the well-known musician Woody Guthrie's nephew, Damon Guthrie, to produce this cartoon depicting much of U.S society rowing together to avert the worst effects of the climate crisis during President Trump and his administration rowing in the opposite direction. I removed the cartoon from the first edition of "In Our Hands" because it was clearly too political, and I wanted the book to be read by individuals of all political persuasions. During the second Trump presidency, he has sought to do everything possible to increase fossil fuel consumption, open federal land for timber harvesting, and gut climate science and public reporting of the issue.

In this book, I feel comfortable including the original cartoon.

CONCERNED
BUT NOT   BUSINESS
NON-PROFITS   ENGAGED   AND THE
AND ENGAGED   CITIZENS   MEDIA   THE TRUMP
MEMBERS   ADMINISTRATION

To win the climate race we have to start rowing together

The future is not, however, preordained. We have a choice as to how this story ends, but to do so, all of humanity must row in the same direction. We need a whole system change to deal with the "economic growth at all costs" and the global warming/climate change crises, and this seems a near-impossible task in the time available. We need to all buy into a new story of humanity that supports the natural world we all depend upon for our survival. This may also require a fundamental shift in consciousness, all inspiring yet daunting challenges.

The gap between the rich and the poor has been increasing – with potentially disastrous consequences.

While the poor in most countries have improved their lot over the years, it is also true that the rich have been getting richer. If the gap gets too big and the poor feel sufficiently suppressed and feel they have nothing to lose, they might overturn the existing order, as was the case in Russia, China, and Cuba.

We are moving from globalization to nationalism, from collaboration to treating the "other" as adversaries.

For most of my life, I have favored building a more integrated, global economic system while supporting fair competition between nations and businesses. However, the world is headed in the opposite direction, towards more nationalism and, in some cases, totalitarianism.

Global collaboration is under threat by President Trump during his second term. He uses fear and power to coerce other nations to do his bidding. In the process, he and the United States are increasingly seen as an adversary and no longer to be trusted. Hopefully, the other countries of the world will push back against this, for it is critical that the nations of the world work together to address such issues as the global warming/climate crisis, global health, nuclear nonproliferation, cyber warfare, and terrorism, just to mention a few global challenges.

United States' leadership in the world is in decline.

Numerous books have been written about the rise and fall of great nations, from the Roman Empire to the waning of Great Britain's influence since World War II. One of the underlying reasons for this is "hubris."

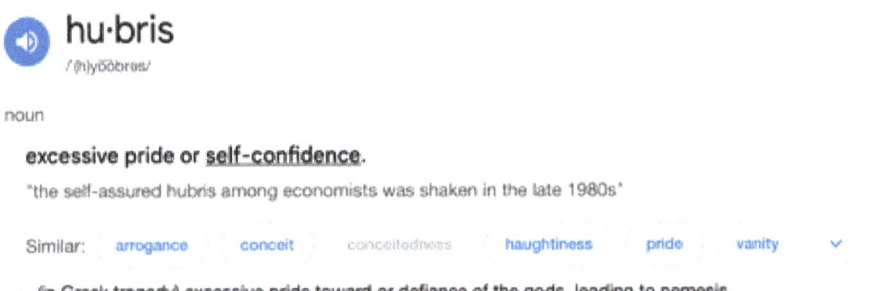

hu·bris

/'(h)yo͞obrəs/

noun

excessive pride or <u>self-confidence</u>.

"the self-assured hubris among economists was shaken in the late 1980s"

Similar: arrogance   conceit   conceitedness   haughtiness   pride   vanity   ⌄

- (in Greek <u>tragedy</u>) excessive pride toward or <u>defiance</u> of the gods, leading to <u>nemesis</u>.

And there is every reason to believe that that fate may someday cause the United States to lose its dominant position economically, militarily, and politically. The United States is no longer respected by much of the rest of the world, as it has been for most of the time since the Second World War. And during the second Trump presidency, we face the prospect that we will be feared more than respected.

The following is a guest opinion piece I sent to the New York Times on April 1, 2025, which states my concerns:

The Clash Between China's Past Humiliation and President Trump's Hubris

President Trump's aggressive efforts to humiliate China, along with many of our allies, are likely to fail for many reasons, China's humiliation by Western colonial powers in the 19th century being central. Some Chinese still refer to this as the "Century of

Humiliation."

This was the result of two "Opium Wars" in the early 1840s when Britain coerced Imperial China to give it access to China's ports and thus control China's international trade. Shanghai became a "treaty port." The United States actively participated and benefitted. Hong Kong became a British colony.

Chinese President Xi Jinping has not forgotten that humiliation and may now decide to regain much of the technological and economic glory China had enjoyed for centuries before the "Century of Humiliation." I believe that this sense of past humiliation is behind China's statement that "China will only engage with the United States based on respect."

China, the world's second-largest economy, has the capacity to deal with short-term pain so long as it sees long-term gain. President Trump, on the other hand, is not as insulated from public opinion as President Xi. Furthermore, the citizens of China are far more willing to accept hardship for the greater good of their country and a restoration of national pride. As a result, "Like bamboo in a strong wind, China will bend but not break."

One possible future: China may keep its tariffs on U.S. imports very high, drop its tariffs on other countries, and encourage nations to trade in the Chinese yuan rather than the U.S. dollar. The United States would go into recession, the dollar would weaken, China's investments in the U.S. treasury bonds, currently at

$760 billion, would fall, and the massive U.S. debt would be further jeopardized.

This would be the turning point in the world's economic and geopolitical history, brought on by President Trump's hubris, defined as "exaggerated pride and self-confidence." And that would be his enduring legacy.

An excellent small book on this subject is "Immoderate Greatness – Why Civilizations Fail" by William Ophuls. Another is "The Rise and Fall of the Great Powers" by Paul Kennedy.

Americans have been moving from "relationships "to "transactions."

When living in small rural communities a hundred years ago, we connected with others at churches and community centers, and family members tended to live near each other. Today, many family members have moved to cities in or outside the United States. We have moved towards a transactional society in which meaningful personal relationships have given way to transactions. There is an emotional cost to this that most people are unaware of because it now feels "normal." Young people use social media to connect, but many find that they are lonely.

We have been shifting from "We" to "Me."

World War II was a period in our history when people focused more on what they wanted to do for their country rather than just for themselves. I fear we have moved far away from that value over the past 80 years since 1945. As President Kennedy put it so eloquently in his inaugural address in 1962: "Ask not what your country can do for you, but what you can do for your country." I recently watched the Netflix series entitled "World War II in Color," during which our military service men and women proudly gave their lives for their country. I fear that that would be much harder to achieve in today's world, especially when the President of our country says of our military men and women who have died fighting for our country: "Why should I go to that (military) cemetery? It's filled with losers."

As I noted earlier, all the observations above express deep concern for our collective future. I therefore would like to end this book with two very positive observations:

Crisis Can Lead to Opportunity.

In 2017, Brookings published a research paper on this topic, often attributed to the Chinese. I include here only a few of their observations:

"During a crisis, incentives and motivations change, potentially leading to new cooperative behaviors and even the creation of new systems or structures. Crises can get the collective adrenaline flowing, focusing minds on solving the problem at hand.

Plato is credited for coining the phrase, "Necessity is the mother of invention." Often, a crisis acts as the forcing mechanism to compel expeditious innovation, leading to rapid advances in technology, policy, and/or procedures.

Large-scale crises that challenge multiple interests and equities have a way of pulling together diverse partners—allies and rivals alike—to solve the crisis. If nurtured, these relationships can then be parlayed into cooperation in other areas.

Global crises that crush existing orders and overturn long-held norms, especially extended, large-scale wars, can pave the way for new systems, structures, and values to emerge and take hold. Without such devastation to existing systems and practices, leaders and populations are generally resistant to major changes and to giving up some of their sovereignty to new organizations or rules.

A crisis has a way of letting the cream rise to the top. In the midst of a situation, those with the right skill sets and talent—even if they are not the identified leaders or top performers—have a way of rising to meet the challenge, creating a dynamic that enables the entire team or group to grow closer and work better together."

Hope can lead to action, and action can lead to hope.

As I have mentioned, I am an optimist by nature, so expressing my major concerns about the future of the United States and the world is very uncomfortable for me. Despite all of my concerns, however, I believe that humanity can rise to the challenges facing us. Recall Winston Churchill's quip, "Americans can always be trusted to do the right thing once all other possibilities have been exhausted."

For some time, I have had a tagline on my emails that read: "Hope leads to action, and action leads to hope." I feel that we must not come from hopelessness, for that defeatist attitude will most certainly kill the impulse to find solutions.

You, me, and the rest of humanity have a choice as to

how this story ends. We are all on a journey, and I hope this book has encouraged you to think deeply about yours. And if you would like to engage with me on anything I have written, I encourage you to send me an email at Wilford@WilfordWelch.com.

Appendix

APPENDIX 1: Assistant Secretary Marshall Green's report immediately following our eleven-nation trip to East Asia in April of 1969

6200

DEPARTMENT OF STATE
Washington, D.C. 20520

April 21, 1969

SECRET - LIMDIS

TO: The Secretary

THROUGH: S/S

FROM: EA - Marshall Green

SUBJECT: A View of East Asia - INFORMATION MEMORANDUM

The attached report on my trip to eleven East Asian countries reaches a number of conclusions based in large part on conversations with top leaders of the area and with American officials:

1.   Mainland China: Peking has never been so extreme and hostile as at present. This is unlikely to change as long as Mao is in control. We should nevertheless liberalize US trade and travel restrictions and make clear our historic continuing friendship for the Chinese people, looking to the post-war period. Asian leaders would understand the political merits of such US moves.

2.   Soviet Role in East Asia: Moscow is in a deep dilemma over how to proceed in the face of an intensifying Sino-Soviet conflict, Indonesia's upheaval, and the costs and risks of supporting Hanoi and Pyongyang aims. Asian leaders believe that Moscow favors a negotiated settlement in Vietnam, but on Hanoi's terms as far as possible.

3.   Vietnam: Our role in Vietnam (and Thailand which some regard as East Asia's pivotal country) is widely supported. Reports of US intentions to Vietnamize the war are welcomed provided this is done in measured fashion. President Thieu believes peace negotiations will reach a conclusive stage six months hence.

SECRET - LIMDIS

MICROFILMED
BY S/S: CMS

Wilford H. Welch

I mentioned to a number of these leaders the probability
that our Government would make moves from time to time designed
to prove that it is Peking not Washington that is isolating
China and to reaffirm our continuing friendship toward the
Chinese people as opposed to the Chicom leadership. Reaction
to this line ranged from equanimity to strong approval. Even
Chiang Kai-shek's son, CCK, responded favorably. He commented
that he realized that the US was a practical nation and recognized from his own long experience in fighting Communist China
that a good political strategy was sometimes just as important
as a good military strategy. No one seemed to share the
Soviet's concern that the US was contemplating normalization
of relations with Peking.

Closely related to this subject is the fate of Taiwan.
Talk of return to the mainland, though muted in recent years,
is nevertheless unfortunate in terms of world opinion, and
scarcely helps in the task of maintaining the GRC's international position. The Republic of China, not Peking, stands
for the traditional values of China; and, in contrast to the
economic stagnation and confusions on the mainland, Taiwan is
entering the big leagues in trade and manufacturing. Here is
where Taipei should be concentrating all its efforts against
Peking. Any reference to return to the mainland associates
Taipei with a forlorn mission which only frightens the ROC's
potential friends and may alienate them. I am certain that
most officials in Taiwan understand this. A notable exception
is the old Gimo.

Soviet Role in East Asia

Most Asian leaders recognize that Hanoi is under con-
flicting pressures from Moscow and Peking with respect to a
negotiated settlement. Few would go so far as did President
Suharto and Foreign Minister Malik in theorizing that Moscow
may be holding back on military deliveries to Hanoi in order
to pressure Hanoi toward a peaceful settlement and that Hanoi
is countering by insisting on military guaranties before
agreeing to do so. Lao Delegate for Foreign Affairs, Khamphan
Panya, agreed with the latter possibility, but both he and
Prince Souvanna Phouma went perhaps even further than Malik in
welcoming the Russians as an important additional counterforce
to China in Southeast Asia.

On the opposite extreme was President Chiang Kai-shek, who maintained that all Communist powers had the same objectives and that Vietnam was no exception. Both he and his son, CCK, argued that Moscow, like Peking, wanted the Vietnam war to drag on and on so as to keep the US occupied.

Most Asian officials disagreed with this thesis. They felt the DRV/NLF struggle is along Maoist lines which Moscow would not wish to see succeed. They saw Moscow as favoring a negotiated peace, though on terms favorable to the DRV/NLF.

This latter view strikes me as valid, being borne out by the record of Moscow's efforts in Paris and elsewhere. I further conclude that Moscow may not have any clear idea as to how to proceed in Asia. Moscow must have been left in a deep dilemma by the widening Sino-Soviet rift, the upheaval in Indonesia, and the costs and risks of supporting Hanoi and Pyongyang in the years ahead.

Vietnam

Leaders of Indonesia, Singapore, Malaysia, and Thailand view Thailand as the pivotal country in Southeast Asia whose security and territorial integrity are vital to all. Since Thailand's security is seen as directly related to the fate of South Vietnam, there is general recognition of the crucial issues at stake in Vietnam. Leaders elsewhere in free East Asia reach much the same conclusion about standing firm in Vietnam, though they may attach less importance to Thailand as the key country.

I heard that there is far less support for our Vietnam involvement amongst Japanese. Although their leaders understand our position and may sympathize with it, they say little in public that is positive. Embassy officials told me that this is the wisest thing for them to do because "high posture" remarks on Vietnam (or Korea) would only touch off a public debate which might weaken the ability of Japan and its leaders to be of help to us in discharging our defense responsibilities in East Asia.

There was general understanding and support for our
efforts in Paris. However, several leaders expressed concern
that the Paris talks would lead to US force reductions in
Vietnam before the ARVN was capable of taking over from the
Americans and before the Government of Saigon was adequately
consolidated. It was their clear preference that the US military presence be retained in Vietnam as long as possible.

Asian leaders strongly approve of Vietnamization of the
war, provided this proceeds at a measured pace. The virtues
of this course of action were seen as several. It would help
buy time with American public opinion which was regarded as
the main target of the enemy's tactics of delay and hold-out.
Vietnamization would also enhance the long-term strength of
the ARVN and the GVN's political stature. All this would prompt
Hanoi, as nothing else would, to negotiate seriously in order
to upset our Vietnamization time-table; but if there were no
settlement by negotiations, we would still be in a better position with American opinion (provided de-Americanization was
carried out fast enough) and the ARVN and GVN would be in a
stronger position.

There were wide differences of opinion as to when the
war in Vietnam would be settled. Two informed views are worth
reporting. President Thieu believed that the Communists are
sufficiently realistic to go for a negotiated settlement while
they have enough strength with which to bargain. They know
that, if they persist too long, there will be little left of
their cadre force with which to seek a take-over after the
peace. He concluded that, after one more offensive, of some
three months' duration, the DRV/NLF would enter into serious
negotiations which would be concluded six months from now. A
top-level American official in Saigon conjectured that serious
talks would start sooner than that, but would continue for many
months marked by Hanoi's fight and talk tactics. I incline to
this latter forecast.

The Mission Team is to be commended on its handling of
the press. It is a well coordinated operation in which our
spokesman provides the facts and leaves it to the press to
interpret those facts. He does not indulge in the predictions
that have caused us so much anguish in times past.

On the other side of the ledger, I was appalled to hear from General Abrams that the three worst ARVN divisions are located in the most critical area, between Saigon and the Cambodian frontier, where enemy strength is being massed. It seemed absolutely incomprehensible to General Abrams that President Thieu, whatever his political reasons, had permitted this situation to persist. General Abrams commented that if the third-rate Generals now commanding these three divisions were replaced by good officers, the effect would be to upgrade significantly each of the divisions concerned. I believe this is an issue of such critical concern to our forces that we must press hard for its correction immediately.

Laos and North Thailand

I found Prime Minister Souvanna Phouma in a gloomy mood over recent military set-backs in Laos. Royal Lao Government forces are tired, their morale low and they are no match on the ground for the 40,000 North Vietnamese who constitute their principal opponent. Because the Government forces failed during the last rainy season (largely through air attacks, which is the RLG's only potent weapon) to recover ground lost during the dry season, and because of increased NVA efforts, the enemy is now in control of more territory than at any time since the 1962 Accords on Laos. The NVA/PL are now poised at several points along the Mekong which has long been regarded as flowing along or through the center of the RLG's zone of influence.

We must identify areas of Laos which are vital to our interests in Southeast Asia and consider what to do if the NVA/PL threatens them. One course of action might be through diplomatic pressures on the Soviets and others to agree on and respect separate RLG and PL zones of influence. Another course of action would be to encourage and assist the Thais, whose interests are most immediately at stake, to take on a more active and extensive military role toward Laos. These two courses could be pursued concurrently. Meanwhile we must continue to provide adequate aid to Laos.

As to areas in North Thailand which border on Laos and whose dissident hillsmen are infected by the PL (as well as by Peking and Hanoi-trained agents), there was some question in

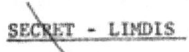

SECRET - LIMDIS

Wilford H. Welch

my mind, after hearing several briefings on the subject,
whether the current RTG policy to root out guerrillas in
remote mountain areas is wise. We start from the fact that
Bangkok's interest and support are inadequate and that the
Thai Commander of the 3rd Army (Forward) is mediocre. Although
we are helping with border police training, the police seem to
lack the aggressive qualities needed for a successful search
and destroy strategy, and the way in which they go after these
hillsmen may create more enemies than they destroy. A policy
of holding the line firmly at the foothills would seem to be
more in line with Thai capabilities and this could spare us
and the RTG a heavy price in funds and casualties. I did not
have time to look into this question and I would defer to the
views of our highly competent Country Team, including its
Chiang Mai component.

The Problem of Divided Countries

During the past fifteen years, there has been striking
progress in economic stabilization and growth, especially
among the more dynamic peoples of free East Asia -- the
Japanese, Koreans and Chinese. There has also been notable
progress in leadership and administration as first generation
revolutionary leaders like Rhee and Sukarno have been replaced
by pragmatists. At one time Communism held forth some attrac-
tion, especially to Asian youth, but it has now been on display
so long that whatever attractions it once held have almost
vanished.

Thus in all these fields -- economic stabilization and
growth, leadership and anti-Communism -- there have been
notable improvements in almost all free nations of East Asia.

But there is one area in which progress has been lacking:
That is in the quest for improved regional mutual security.
Here the United States continues to bear the overwhelming burden.

A basic reason for this phenomenon is that four of the
world's five divided countries are to be found in East Asia --
Korea, Vietnam, Laos and China. These countries are focal
points of world tensions as the Communist areas attempt to
take over the free areas. Small powers in East Asia seek to
stay out of these struggles, and larger powers like Japan are
strongly averse to involvement. Non-aligned countries such as

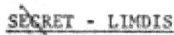

SECRET - LIMDIS

Indonesia, Burma, and Cambodia not only remain aloof but are
even hesitant to join organizations which include divided
countries.

The Quest for Greater Mutual Security

All this makes the road to mutual security more difficult and the prospect for improved military cooperation bleak.

In fact, such prospects in Northeast Asia are growing
less rather than greater, as Japan's urban sprawl engulfs our
bases and Japan seeks its own identity. I anticipate we will
be forced to restrict our base presence in Japan several years
hence to Yokosuka, Sasebo, Misawa, and possibly Yokota. We
cannot and should not resist Okinawan reversion; we can only
seek the best terms possible for operating our bases there.

Moreover, Japanese public opinion does not focus on
the real issues at stake in defense of the Republic of Korea,
but concentrates on such peripheral issues as the dangers of
Japan's becoming embroiled in Korean hostilities through our
use of Japanese air bases. As has been demonstrated by the
Blue House Raid, the Pueblo incident, and the EC-121 shoot-
down, Pyongyang is seeking to play on US public war weariness
and to exploit rifts between Seoul, Tokyo, and Washington.
I believe our next Policy Planning meeting with Japanese
counterparts should concentrate on this subject.

In this situation, the Republic of Korea seeks firmer
assurances and support. It looks for a wider regional
security association in the form of PATO, but this move seems
foredoomed since the only takers are Taipei and Saigon.
Even Thailand is uninterested; and the US, whose involvement
will be essential if PATO is to have any real meaning, has
expressed coolness toward any new commitments.

Foreign Ministers Thanat and Romulo spoke to me of the
need for alternatives to SEATO. I replied that we must make
the best of SEATO, whatever its many defects, for there is no
other practicable alternative at this time to provide a US
commitment to defend Thailand.

SECRET - LIM DIS

APPENDIX 2: Timeline of the U.S. steps for Henry Kissinger's meetings with China

TOP SECRET/SENSITIVE/EXCLUSIVELY EYES ONLY

Direct and Indirect Specific Messages Between
The U.S. and PRC

October 10, 1969. Sher Ali Khan tells HAK that Chinese have been told that Yahya is ready to talk about US intentions in Asia when Chou visits Pakistan. Sher asks HAK for more specific message for Chinese, is told by HAK that Yahya might say US is removing two destroyers from Formosa Straits. (Tab A)

December 19, 1969. Hilaly tells HAK that in early November Yahya had told Chinese Ambassador of US interest in normalizing relations and of the two destroyers. Peking responded that it appreciated Pakistan's role and had released two Americans (yachtsmen). HAK says he would give Pakistan more to say to Chinese when Chou visits there, and the President is ready to establish a secure channel. (Tab B)

December 23, 1969. Hilaly tells HAK that further letter from Yahya says that Chinese appear willing to resume Warsaw talks without preconditions and that they are worried about US-Japanese agreement and Japanese militarism. (Tab C)

February 11, 1970. HAK gives text to Derksen in meeting with him and Lodge. The message says that the President is ready to establish a more secure channel than Warsaw for matters of the most extreme sensitivity, either through Derksen or General Walters. (Tab D)

February 22, 1970. Hilaly relays to HAK Yahya assessment of Chinese thinking about US-Chinese relations (no explanation whether this based on further contacts with Chinese). Yahya says US initiatives have encouraged Chinese, who no longer see US-Soviet collusion and emphasize that US should not construe Chinese readiness for meaningful dialogue as a sign of weakness. Chinese response likely to be measured, but they seem inclined for meaningful dialogue on all issues. China-US war now seems very remote, with possibility of Vietnam war expansion having lessened. HAK says Yahya should tell Chinese that we can't control press speculation and that we are ready for direct White House channel to Peking. (Tab E)

TOP SECRET/SENSITIVE/EXCLUSIVELY EYES ONLY

TOP SECRET/SENSITIVE/EXCLUSIVELY EYES ONLY   -2-

May 3, 1970. Winter comes to see HAK to tell him of the Chinese
reaction to the Cambodian operations. The Chinese had asked him
whom they could talk to and trust. HAK gives him a message for the
Chinese saying we have no aggressive intentions and would like to
establish regular relations. HAK is prepared to talk to a person of
stature on the Communist Chinese side if this can be done secretly.
The Chinese can reply by getting in touch with General Walters. (Tab F)

June 15, 1970. HAK gives message to Walters to use as talking points
with his Chinese contact in Paris, suggesting a channel through Walters
and our readiness to send a high-level personal representative to the
President to Paris or elsewhere for direct talks on US-Chinese relations. The message notes the publicity surrounding the Warsaw talks
and suggests this alternative channel "for matters of the most extreme
sensitivity" with knowledge of the talks "confined to the President, his
personal advisors and his personal representative unless otherwise
agreed." The purpose would be "to bring about an improvement in
US-Chinese relations fully recognizing differences in ideology." (Tab G)

October 25, 1970. President tells Yahya in Washington meeting that
he should tell Chinese when he visits Peking that US-Chinese dialogue
is essential, we would make no condominium against them, and we are
ready for high level discussions. The President and HAK emphasize
that we are talking not of a hot line but a willingness to send senior
person, such as HAK, to mutually convenient capital for serious talks
with absolute discretion. (Tab H)

December 9, 1970. Hilaly tells HAK of Chou's reply to President's
oral message through Yahya. Chou's reply, coordinated with Mao and
Lin Piao, cites Taiwan issue and says that "in order to discuss this
subject of [vacating] Chinese territories called Taiwan, a special envoy
of President Nixon's will be most welcome in Peking." Chou notes
previous US messages but that this is first at Head of State level. The
Chinese attach importance to this message since the US knows Pakistan
is a great friend of China. (Tab I)

December 16, 1970. Winter tells Jon Howe in HAK's absence that the
Chinese had asked on December 8 for names and data on persons they
could invite to China who would be friends. (Tab J)

TOP SECRET/SENSITIVE/EXCLUSIVELY EYES ONLY

December 16, 1970. HAK gives Hilaly memorandum for Yahya which states US pleasure at Peking's offer proffered at the February 20 Warsaw meeting to receive a US representative to discuss outstanding issues between our two governments; welcomes high level discussions seeking the improvement of relations between our two countries (not limited to Taiwan question); and proposes a meeting of our respective representatives at the earliest possible moment to discuss the modalities of a higher level meeting in Peking. (Tab K)

January 11, 1971. Ambassador Bogdan tells HAK that Ceausescu sent his Vice Premier to Peking (and Hanoi) after his conversation with President Nixon. The Vice Premier had extensive talks with Chou En-lai who handed him a message (which refers indirectly to the communication we had sent through Pakistani President Yahya). This message says that Taiwan is the "only one outstanding issue" between the US and China, and if we have a desire to settle it and a proposal for its solution, the PRC would be prepared to receive a special US envoy in Peking. Chou En-lai added the comment that since Nixon had visited Bucharest and Belgrade, he would also be welcome in Peking. (Tab L)

January 12, 1971. Sainteny letter (through Smyser) relates that he saw the Chinese Ambassador December 23, 1970 and passed the word that we were looking for a channel. The Ambassador "received the idea with a certain reserve" but transmitted it to Peking. (Tab M)

April 27, 1971. Haig (for HAK) sends package to Walters via courier to Paris. Letter to Walters encloses HAK to Sainteny letter and note for Chinese. Walters is to contact Sainteny and show him HAK letter which asks for him to arrange meeting between Walters and Chinese Ambassador in Paris (or other senior official). At such a meeting Walters is to deliver note to Chinese which offers to establish a secure channel and send HAK to Paris for direct talks with whomever PRC designates. (Tab N)

Appendix 3: Chapter 5 of the 2021 edition of "In Our Hands, A Handbook for Intergenerational Actions to Solve the Climate Crisis," by Wilford H. Welch.

THE ROAD TO RUIN - HOW OUR ACTIONS AND INACTIONS LED TO OUR COLLAPSE

This view of the future may seem far-fetched to many of you, but I am convinced that it is certainly possible. This scenario assumes there is little change in the world's rapacious use of fossil fuels. It is indeed a worst-case scenario, which I pray will never happen, but it is included here as a reminder of what could happen if we do not wake up and take action this decade. Only if the world makes immediate, dramatic reductions in our greenhouse gas emissions, and in many of our current actions, can we avoid many of the outcomes described here.

Looking back from the year 2050

In this future, humanity heads down a path toward ecological disaster as well as economic and societal collapse.

Some of the forces that caused this tragedy were:

• Lack of the individual, collective, and political will to take the actions that were clearly called for and within our reach.

•Nations inciting fear of "the other," making cooperation on a global scale impossible.

• Reluctance to get off of fossil fuels fast enough.

• Tipping points and irreversible feedback loops reached.

The 29 years between 2021 and 2050 have been the most turbulent and destructive in human history. Our missteps in some areas and inaction in others have led to our near collapse today, defined by the following:

The world's temperature is now 7.0 degrees Fahrenheit above what it was in 2020. While some parts of the world in 2021 were making major efforts to switch from fossil fuels to renewable sources of energy, most countries failed to meet the commitments they had agreed to at the 2015 Paris Climate Summit and the UN Climate Change Conference in November 2021 in Glasgow, Scotland. What they did do was "too little, too late."

By the late-2020s, rising temperatures had caused much of the Arctic to be ice-free for much of the year. It also caused much of the tundra to thaw, resulting in millions of tons of methane being released into the atmosphere, causing yet more global warming.

With much less ice and snow in the Arctic, the sunlight that had previously been reflected away from Earth by the snow was absorbed by the darker surface of the ocean, not only making the ocean warmer but also making it more difficult for the ocean to absorb as much CO_2 as had been the case for millions of years.

The median level of the oceans is above where it was in 2020. Melting ice from Greenland, Antarctica, and, to a lesser extent, the Arctic (because it is not a landmass but open water) has raised sea levels a couple of feet. This has been high enough to flood portions of many coastal cities around the world. Miami, New Orleans, Boston, New York, Amsterdam, Shanghai, and Venice are now among the many coastal cities impacted. Most experts expect the sea to rise another three feet by the end of the century.

Drought has destroyed agriculture and livelihoods. Millions of people in the Middle East, Africa, and South Asia are experiencing desertification of their farmlands, forcing them to become climate refugees.

Temperatures in excess of 120 degrees Fahrenheit are commonplace around the world, including parts of the United States. Fires have consumed huge forests and some towns. Much of the Amazon rainforest, "the lungs of the planet," has dried up, in part due to continued logging and land clearing for soybean farming and beef production.

The snowpack and glaciers of the Himalayan mountains that fed the great rivers of China, India, and Southeast Asia for centuries have melted. This melt initially caused flooding, but now, many of these downstream communities are facing insufficient amounts of fresh water for farming.

The glaciers of the South American Andes, where melting snow had for millennia supplied 80 percent of the freshwater to downstream populations, have also been reduced to a fraction of their former size, with equally devastating consequences.

In the American Southwest, conflicts over fresh water have become frequent and intense as farmers, urban communities, developers, and natural gas frackers battle for access to ever-diminishing quantities of groundwater and underground aquifers.

The world's oceans have lost 90% of their marine life because of massive factory fishing operations and acidification. Our oceans, overfished and a dumping ground for plastics and toxins, are turning into acidic "dead zones." Massive factory fishing fleets, whose drag nets scraped the bottom of our oceans and seas for decades, ensnared virtually every fish in their path and have emptied our oceans of most marine life.

And, because the planet's oceans, making up 71 percent of the Earth's surface and 97 percent of the Earth's water, have grown warmer and warmer, they can no longer absorb as much CO_2 as in previous centuries, and their acidity has risen dramatically.

Ninety percent of coral reefs around the world are now dead and no longer provide habitat for spawning fish and the fish that humans had depended upon for their food.

More than 80 percent of all species of animals alive in the year 2000 are now extinct. Climate change, overpopulation, and pollution have destroyed habitats, and poachers have killed off our most iconic mammals in the wild. Lions, tigers, elephants, bears, dolphins, and big apes exist mainly in a few zoos and "wildlife zones."

Climate refugees have no place to go and have died by the millions. In the 2020s and 2030s, the United States, Western Europe, and other developed countries closed their borders to climate refugees. Now, in 2050, climate refugees from the Middle East, Africa, South Asia, and South America, facing ever more dire circumstances in their own countries, are roaming the planet.

Somalian refugees with little food and water and no place to go

Image courtesy of Jerome Delay/Associated Press

International trade and investment are now a fraction of their previous size. Starting in the 2025s, when the US erected tariff and non-tariff barriers to international trade, investment, and technology flows, other countries responded in kind in an effort to stimulate their own economies and protect jobs. Ultimately, all this disruption proved disastrous for individual countries and the global economy. The prices of commodities and consumer goods shot up in the US and other advanced economies, which had been importing from lower-cost countries. The economies of the lower-cost countries declined because of their loss of export markets.

The US economy is severely stressed, and the global economy is on the verge of collapse. Seriously stressed by ever-more-costly climate disasters, sea-level rise in major urban areas, resource shortages, agricultural failures, reduced international trade, and social unrest, the U.S. economy has become more and more unstable.

Interest payments on the massive amounts of debt that had long sustained economic growth cannot be maintained in the U.S. or globally.

The "politics of fear" carry the day as demagogues and authoritarian power structures have gained power everywhere.

Economic and social calamities have caused people to react out of fear and scarcity. Even in previously strong democracies, ruthless politicians have preyed on their people's mistrust of "the other." The global political environment has become ever more toxic and debilitated by short-term decision-making and the vested interests holding on for dear life. Vast economic inequality threatens the very fabric of life.

The divide between the very few who can afford to live in fortress-like "gated communities" and the billions who now live in poverty has taken its toll throughout the world. The "haves" seek protection with private armies. Some cities are armed camps, and terrorist actions are everyday occurrences.

In sum, this is life in 2050. Warnings raised in the 2015 book "Overdevelopment, Overpopulation, Overshoot" have all come to pass. Humanity was not up to the task of addressing global warming and climate change before it was too late. Humanity proved incapable of moving beyond its adolescent phase to an era in which collaboration with other human beings and harmony with the natural world would have saved us.

www.ingramcontent.com/pod-product-compliance
Lightning Source LLC
Chambersburg PA
CBHW051620120626
46551CB00014B/1870